MIX
Papier aus verantwortungsvollen Quellen
Paper from responsible sources
FSC® C105338

If you have any concerns about our products,
you can contact us on
ProductSafety@springernature.com

In case Publisher is established outside the EU,
the EU authorized representative is:
Springer Nature Customer Service Center GmbH
Europaplatz 3, 69115 Heidelberg, Germany

Printed by Libri Plureos GmbH
in Hamburg, Germany

Economics of Renewable Energy

Yoram Krozer

Economics of Renewable Energy

An Assessment of Innovations with Statistical Data

Springer

Yoram Krozer
Director Sustainable Innovations Academy
Amsterdam, The Netherlands

Visiting professor at the Graz University of Technology
Graz, Austria

Emeritus professor at the University of Twente
Enschade, The Netherlands

ISBN 978-3-030-90803-4 ISBN 978-3-030-90804-1 (eBook)
https://doi.org/10.1007/978-3-030-90804-1

This Springer imprint is published by the registered company Springer Nature Switzerland AG
The registered company address is: Gewerbestrasse 11, 6330 Cham, Switzerland

Preface

This book, *Economics of Renewable Energy* is about socio-economic mechanisms of change that can generate accessible, clean energy production and consumption. While the present avalanche of publications on renewable energy focuses on the future of energy and climate, this book reflects on changes in the past, mainly with statistical data used for nearly 60 graphs, tables, and figures in the main text. The mechanisms that enhance or impede changes toward renewable energy are identified, which is instrumental for decision makers, policymakers, entrepreneurs, scholars, and students that are interested in energy. Experts and laymen that read the draft manuscript found it information dense, while well readable and interesting. For basic data, please contact the author by mail (krozer@xs4all.nl).

The idea for writing a book focused on the mechanisms of change emerged a few years ago, after my involvement in business start-ups, citizens' initiatives, and policies on local, regional, and national scales. Along with enthusiasm about renewable energy, the sluggish introduction is observed. Why renewable energy is adhered, but changes move slowly? Why do many innovators pursue renewable energy, but only a few have gained from their efforts? Why do firms reluctantly change their course of action despite growing markets in renewable energy? Why do policies aim to foster renewable energy but support interests vested in the past? Above all, what mechanisms can generate changes in energy production and consumption for broad welfare in communities and countries? Answers are based on my interpretation of the statistical data using theories and experiences. Theories are found in literature while experiences are gained by co-operation with other people.

It has been my good fortune to get help from many wonderful people. The joint national plan of trade unions and environmental groups for energy and environment in the mid-1980s was a great experience in societal breakthroughs, thanks to Hans Becht, Jacqueline Cramer, Ferd Crone, Cor Inja, Andries Nentjes, Johan Stekelenburg, and Lucas Reijnders. Kornelis Blok ICARUS model in the early 1990s was an eye-opener to the economic modelling of renewable energy, which is used in the DESC model for Unilever, thanks to Jeroen Bordewijk, Rob Donia, and Cees van Leeuwen of Unilever, and was subsequently applied for the energy issues at Statoil, thanks to Fredde Cappelen and Lars Sund; at Shell, thanks to Henk Groeneveld; and in other energy companies during the early 1990s. I gained hands-on experience in the first 'energy-neutral' building in the Netherlands – Pelgromhof in Zevenaar in the mid-1990s – thanks to Frans van de Werff, Han Brezet, and Pascale van Duijsse. During the 2000s, the Cartesius Institute enabled to experience innovators in the Frisian Solar Challenge, thanks to Wubbo Ockels, Bouwe de Boer, and Andries van Weperen; incumbent interests in Energy Valley, thanks to Hans Alders and Gerrit van Werven; local energy projects, thanks to Stephan Jansen and Harm-Jan Bouwers; as well as local energy policies, thanks to Nienk Hoepman and Simon Tijsma. Pauline Westendorp, Frank Boon, and Aukje van Bezeij were excellent guides in cooperative engagement. It was my honour to be involved in the Arab Renewable Energy Commission (AREC) by His Royal Highness Asem bin Nayef, Her Royal Highness Sana Asem, and Mohammed Al-Taani in Jordan, as well as by Yelena Shevchenko in the Kazakhstan R&D policy on climate change, and by Boglarka Vajda in the Romanian Green Energy Business Cluster. I started to grasp the evolutionary economics due to Maarten Arentsen of CSTM and the perspective of chemical engineers thanks to Michael Narodoslawsky of TU Graz. All those experiences were instrumental for assessing the mechanisms of change in renewable energy.

I am grateful for the cooperation of international trailblazers in the communities' renewable energy: Relinde Baeten and Dirk Vansintjan of Ecopower in Belgium, Søren Hermansen of Samsø Energy Academy in Denmark, and Begoña Urien Angulo from the province of Navarra in Spain. Comments by Mohammed Al-Taani, Maarten Arentsen, and Lian Staal are highly appreciated.

Last but not least, I am thankful to Shonali Chenzira from India for editing this book. I also highly appreciate the cooperation with Springer Nature, in particular Margaret Deignan, Deepthi Vasudevan, and Vigenswaran Balachandrane. Obviously, I bear all responsibilities for shortcomings.

This book is dedicated to a bright future for my daughter, Mira Krozer.

Amsterdam, The Netherlands — Yoram Krozer

Contents

Chapter 1
Introduction

What are the possibilities to generate a shift from fossil fuels to renewable energy? This book discusses the socio-economic mechanisms that enhance renewable energy in businesses and households. That discussion is largely based on the interpretation of statistical data on energy resources and pollution, historical changes in energy technologies and services, innovations in the energy markets, consumption growth of renewable energy, as well as the decarbonisation and valorisation trends across countries. All that data is public, but the interpretation is the author's responsibility. In this introductory chapter, the aims and assumptions, theoretical perspectives on innovations, data and methods, and contents of the book are presented.

1.1 Aims and Scope

The question of what socio-economic mechanisms drive changes in energy production and consumption is discussed with regard to the possibilities of the global shifts in energy resources from fossil fuels to renewable energy within a few decades, often branded 'energy transition'. The term 'shift' is considered in the sense capturing of the largest share in the global energy market. Mechanisms of change in societies are searched based on experiences rather than ideal policies in theory that usually entail disappointing practices. Those mechanisms refer to the tangible interests in novel technology and behavioural changes; in particular, to energy producers that exploit energy resources for deliveries of products that enable heating and power, activities that convert those products into various energy services, and consumers striving to obtain valuable qualities for energy consumption. The possibilities and impediments of those shifts are assessed mainly through the interpretations of statistical data on energy production and consumption throughout the last centuries and decades. It focuses on tangible results rather than on intentions, ideologies, and institutions because decisions about new technologies and behaviour are usually less contentious than about politics and institutions.

Y. Krozer, *Economics of Renewable Energy*,
https://doi.org/10.1007/978-3-030-90804-1_1

The mechanisms of change are considered within a normative framework that refers to the individual freedom for production and consumption, taking into consideration impacts on various interests, which requires decisions that respect individual choices combined within societies (Sen, 2002). This idea about freedom implies the decision-making aiming at welfare in the broad sense of satisfying individual and social demands, and aspirations for income, wealth, leisure, care, and other values based on decision-making across generations, sexes, and races; all that given scarce resources (Sen, 2009). Such decisions are made by individuals in transactions about private goods, as well as by collectives about common goods. Whilst the former decisions are usually based on the prices of private goods, the common goods are rarely priced. The deficient pricing is because either the decisions refer to love, friendship, care, nature, and other valuables addressing 'moral goods'; considered priceless, or because they cause negative impacts on interests not involved in those transactions, thereby nobody wants to pay for, called the 'external effects' (Tirole, 2017). For example, neither energy access is fully priced, nor emissions of carbon dioxide (CO_2 emissions) from combusting fossil fuels.

For the interpretations of statistical data, it is assumed that people possess nearly unlimited capabilities in using capital, labour, and knowledge resources for the generation of energy resources and converting them into pressure, movement, electricity, light, sound, vibration, and other qualities in energy services which are appreciated and valued in consumption. A challenge is access to those services regarding the uneven distribution of income, capabilities, and other conditions for energy consumption. Moreover, conversions of material and energy resources pose limitations because each activity causes dissipation of energy and dispersion of materials, called emissions, which pollute the environment. As the consumer demands for the valuable qualities in energy services grow emissions can also increase, which may undermine welfare. The problem is how to increase those energy services in order to satisfy the demands of all people, and yet to prevent pollution. Attention to renewable energy is motivated by its availability and low emissions. These resources are found available all over the Earth in useful quantities but they are variable, low-density quantities, thereby need space, can cause a nuisance, and disturb nature.

The wide availability of renewable energy enables local access to energy, which is important for 2.8 billion people that miss electricity at households (IEA, 2020); amounting to a third of the global population in energy poverty. Meanwhile, energy poverty is covered in many publications; for example, from the ethical perspective (Guruswamy, 2011), political one (Sovacool, 2012), consumers' viewpoint (Jessel et al., 2019), and others. It is also considered in policies. However, when policies reduce energy poverty through subsidies for energy resources, they render energy savings inefficient, cannot spend the equivalent of subsidies in support of low income rather than cheaper energy and generate more pollution when fossil fuels are supported. In effect, such policies render wasteful energy consumption and impede efficient energy services. Local renewable energy is an alternative for energy poverty. This possibility is addressed with regard to its availability on all locations on Earth.

Much scholarly and policy attention is also given to emission reduction of CO_2 because enables mitigation of climate change. Low CO_2 emissions on the global scale are envisioned in assessments commissioned by the World Wildlife Fund (Deng et al., 2012), Greenpeace (Teske et al., 2015), and the International Renewable Energy Agency (Gielen et al., 2019). The possibilities are also illustrated in the studies on the United States of America (USA), the European Union of 28 member countries (EU), Indonesia, Malaysia, Western Balkans, and other areas, industries, mobilities, and households. All those scenarios underpin the technological capabilities in the mitigation of climate change. However, they do not resolve the issue about how these capabilities can be employed. In those assessments, it is usually assumed that policies decide about the reduction of CO_2 emissions through energy-saving and renewable energy, and implement actions in accordance with that decision-making. Indeed, several countries announced ambitious plans to attain nil CO_2 emissions in energy consumption within a few decades; for example, the Republic of China (China), EU, Republic of Korea (Korea), United Arab Emirates (UEA), and United States of America (USA). However, not many policies walk the talk as they support fossil fuels in India various ways which cause the growing CO_2 emissions and increases uncertainties about future changes. Market and community-based initiatives can also drive changes with the support of policies. These possibilities are also addressed with regard to valuable by energy services based on renewable energy.

Herewith is focused on the shifts from fossil fuels based on coal, oil, natural gas, and nuclear materials toward the traditional renewable energy resources that are produced for many centuries as biomass for heat and hydro for power, as well as the modern ones produced for decades as geothermal and biofuels for heat, as well as wind, solar and wave for power. Fossil fuels are extracted from stocks, whereas renewable energy refers to the exploitation of energy flows. This distinction is relevant because the stocks can be depreciated, whereas the flows of renewable energy depend on the environmental conditions that pose limits for the deployment of renewable energy as they vary over time and per location. The energy density of renewable energy is also low compared to fossil fuels that have been fossilised over millions of years in Earth. So far, traditional renewable energy and fossil fuels are used for the based load of energy, which means for continuous supplies, whereas modern renewable energy is mainly used for the peak loads of electricity and heat. That distinction within renewable energy is relevant as the traditional renewable resources are mainly used in low-income countries, where the modern ones are hardly applied. Note that some scholars address low-carbon technologies when including nuclear resources with regards to low CO_2 emission, but these resources are fossil fuels.

Energy technologies are considered in value chains, meaning subsequent value-adding activities from energy production toward consumption. Stocks of fossil fuels are processed for the generation of heat and power; for example, coal is processed into cokes, oil in petrol, and other energy products. Those products are usually combusted for heat and power that are distributed through networks called grid, or for local consumption called off-grid. The flows of renewable energy are often consumed directly (off-grid), which implies no distribution and negligible cost.

For example, solar energy inflow delivers light and warmth, biomass provides shade, wind and water flows create cooling, geothermal energy enables cooking and healing. Although they enable life, these passive uses of renewable energy are largely neglected in energy engineering and economics, as well as in this book. Those flows are also converted into denser resources of heat and electricity without combustion. Produced heat and electricity with fossil fuels and renewable energy are also converted for storage as chemical energy in fluids, rocks, batteries, and hydrogen, as well as mechanical energy mainly in pumped water and air. Electricity is a product that is consumed for power or converted for air conditioning, lasers, radars, and other services. It means that the value chains can involve many conversions before consumption which increase the value added by energy services.

Why a new book is presented whilst possibilities of the energy transition are shown in many reviews; for example, from the perspective of engineering (Ghosh & Prelas, 2009, 2011), business interests (Asplund, 2008; Siegel, 2008), historical development (Smil, 2017), policymaking (MacKay, 2009), consumers' viewpoint (Quaschning, 2008); the economic perspectives can also be found in several books and journals. In this book, the emergence and growth of renewable energy is explained as the innovation spur toward the value addition by energy services, thereby a trend that generates welfare because reduces costs and adds beneficial qualities which can be fostered or obstructed by policies. This is based on the observations of innovations in the past centuries and decades. Innovations are comprehended as 'doing things differently' (Schumpeter, 1939, 1989:59) due to novel qualities of processes, products, services, designs, images, models, and other technologies aiming at profits, meaning higher income than all costs and risks. They emerge due to the entrepreneurial initiatives of individuals in firms and communities, and disseminate when novel qualities of technologies are purchased by subsequent customers in the value chain; qualities refer to attractive properties for customers which can be clean and silent energy production as an example. When innovations reduce costs of resources or increase the value of products and services higher productivity is generated, meaning higher output per resource input measured by money. This renders the income growth and generates welfare if benefits of productivity are properly distributed (Helpman, 2004). Innovations that intentionally contribute to the maintaining availability of private goods and common goods are considered as 'sustainable innovations'; herewith, renewable energy is considered an innovation that contributes to the Sustainability Development Goals of the United Nations (UN, 2020).

The focus on mechanisms of change in this book causes two important limitations. One is that several energy resources are not covered. Links between energy and food are ignored, though food provides energy for labour of animals and people. Furthermore, illegal and informal distributions of fossil fuels, biomass, peat, and other resources are also omitted. Nonetheless, these activities are large-scale in many countries. In Nigeria, as an example, annually nearly 7.4 million tonnes of oil is stolen for domestic sales and in the neighbouring countries, which constitutes the market value of USD 5 billion (IEA, 2014). The second limitation is that the impacts are briefly touched. The environmental impacts of fossil fuels are hardly covered, aside of climate change. Impact of renewable energy is not addressed, albeit relevant. For example, the consumption of minerals is neglected but the renewable energy technologies consume twice to four times more mineral mass per energy unit than fossil fuels which need recycling (IEA materials, 2021). Societal debates about renewable energy are also neglected. For example, do biofuels reduce or increase environmental impacts? Is hydropower a blessing or disguise? Does geothermal energy pollute soil and water underground? Are windmills acceptable elements in nature and landscape? Will solar energy increase pressures on land? Answers to such questions require societal debate. Science can provide a basic understanding of the issues and tools for the structuring of these debates. Finally, the issues related to employment, human capabilities, and education are largely neglected as they are regularly covered by the publications of the International Renewable Energy Agency (e.g., IRENA Jobs, 2020).

1.2 Innovation Perspective

Rather than assessing the economic fundamentals that drive changes in energy, which is addressed in other works (e.g., Byrne et al., 2014), or searching for a new innovation theory, the available economic theories on innovations are used for the interpretation of statistical data. The main trains of thought are presented below, focused on the possibility of generating larger incomes alongside less pollution. This possibility has been debated for centuries. Without going into details of that debate, the pessimistic opinions that collide with the optimistic ones are introduced. For convenience, they are marked as the 'environmentalist' and 'economist' perspectives. These colliding viewpoints are briefly introduced, followed by an explanation of basic concepts in the innovation theories, and their viewpoints on the drivers of innovation in energy.

From the environmentalist perspective, economies act within the closed system of Earth, meaning without external inflows of energy and materials. On Earth, space, fossil fuels and minerals are considered scarce, and non-reproducible resources, contrary to reproducible bioresources, labour, capital, and knowledge. In this line of reasoning, when populations expand to the limits of scarce, and non-reproducible natural resources, economies collapse because consumption eventually surpasses these limits, termed 'carrying capacity' of the environment. This environmentalist thinking, derived from the population dynamics, refers to the argumentation based on observations of travellers to many countries during the 1700s. In this argumentation, societies are trapped into recurring cycles of expanding population beyond fertility of the land, because people multiply fast when their incomes grow, followed by hunger and mass starvation when the growing population approaches the carrying capacity of the land. Therefore, the policy of restrain in income and fertility was recommended (Malthus, 1826).

This recommendation was embraced by many opinion leaders and politicians of England in the 1800s, which remained in the Anglo-Saxon political tradition of North-West Europe and the USA throughout the last two centuries. By the 1960s, it was rephrased as the ‘Spaceship Earth’ (Boulding, 1966), followed by arguments about surpassing limits in biodiversity, climate, toxicity, and other environmental issues on the global scale (Röckstrom et al., 2009). That political tradition is also reflected in the economic thinking that advocates limits for income growth; for example, by a global think-tank called ‘Club of Rome’ in the 1970s (Meadows et al., 1972), a steady-state economy with the distribution of income and nil growth during the 1990s (Daly & Cobb, 1994), and the decreasing economies’ ‘degrowth’ in the 2000s (Jackson, 2011).

Although the degradation of environmental qualities occurs beyond a doubt, the idea of lower income is counterproductive because the consumption shifts to basic goods that are usually material and energy-intensive. The consequences are larger consumption of energy, materials and pollution per individual income along with stronger social resistance to the environmentalist cause. These obstacles to the improvements of environmental qualities are observed during economic crises. An alternative, the economist idea is moderate income growth, combined with income distribution and pollution reduction as a result of low-impact technologies, called ‘sustainable development’. This is comprehended as a process of changes toward higher value and lower impacts, resulting from a combination of lower energy losses, the substitution of energy-intensive production, shifts from fossil fuels to renewable energy, and suchlike innovations. This can go on because the Earth is an open system that generates huge solar energy from space as irradiation; it also obtains materials from space but this 5×10^3 tonnes is negligible in nearly 6×10^{21} tonne materials on Earth (Rojas et al., 2021). Materials do not vanish from Earth. They disperse, and can be recycled; referred to as the ‘circular economy’. Energy dissipates in every conversion of resources, which enlarges entropy, or chaos, which is countered by the inflows of solar and geothermal energy (Georgescu-Roegen, 1971). Based on these observations, it can be concluded that processes in nature and economies can sustains long as the capture and storage of energy inflows exceed the dissipation of energy.

In the economist argumentation, key human capability is the development of tools that enable substitutions of scarce resources for more abundant ones; for example, fossil fuels for renewable energy. Therefore, incomes can grow as long as non-reproducible resources are replaced by more abundant, reproducible ones (Solow, 1973). The main challenge to the income growth is pollution because of emissions into the environment whereas the environmental sinks for pollution are limited; in this economic argumentation the discharges are the unintended effects of transactions on other interests, it is the ‘external effect’. Pollution can be prevented through technologies for emissions reduction when policies put a price on those discharges with regard to the scarce sinks (Kuipers & Nentjes, 1973). In this economic thought, imperfections are observed and can be countered by innovations in technologies that drive changes in production and consumption, if the external effects are priced. Opinions about those human capabilities vary from the wholehearted trust in science and technology for resolving social and environmental issues with regard to past trends (Pinker, 2018), to the tedious manoeuvring between the degradation of environmental qualities and well-being for all people (Raworth, 2017).

Usually, environmentalists and economists consider pollution as an event external to the economic activities that need to be resolved by lower volume and income, regulatory authorities or agreements between the interests of polluters and the affected. When proper regulations are enforced or payments are set, the pollution problem is resolved as if it is done automatically. This argumentation is possible because environmentalists and economists consider knowledge about pollution reduction as a 'black box'. An often-used metaphor in economics is that people do not need to know about the working of engines in order to drive a vehicle, though it is handy to comprehend some basics in order to avoid damages. A basic feature is that pollution depends on knowledge about how to use material and energy for targeted products thereby preventing the deficient products. The targeted products are valuable because they are demanded, meaning that people are ready to pay for them. It also implies that non-demanded products must be prevented. This is needed, first of all, because costly resources are wasted. If the non-demanded products unintentionally emerge, they can be reworked with additional materials and energy. Secondly, when the non-demanded products emerge without rework but discharge into the environment because this is cheaper than the rework, these discharges dissipate as emissions which cause pollution. That knowledge about using materials and energy for the generation of products on demand is called know-how.

The role of know-how in conventional economic thinking is shown in a model for loss prevention. The model shows a production, function of material resources without any know-how, but solely as experimentations by trial and error; this is similar to cooking with new ingredients. In such production, resources (inputs) are converted into qualities (outputs), being factorial functions of inputs. It is formally:

$$N_i = N_o! + 1 \tag{1.1}$$

For 'N_o', number of qualities (outputs); 'N_i', number of resources (inputs).

With three resources, seven possible qualities must be considered (3*2*1 + 1 = 7); but the occurrence of six qualities must be averted in order to obtain solely the demanded product assumed superior by the producer. With four resources, 25 qualities are possible (4*3*2*1 + 1 = 25) and 24 must be prevented so as to obtain the demanded product; it is and so on (van Leeuwen, 1989).

Given that even the simplest production involves several material and energy resources in order to deliver a demanded product, huge losses are generated without sufficient know-how. Larger know-how reduces the losses, which implies more valuable products and fewer losses. More complex conversions when more resources are involved, require greater know-how. Moreover, in some conversions it is difficult to contain the losses; for example, in explosions when events evolve very fast because driven by dense energy resources. Although the losses are signals of deficiencies, emissions into the environment are not contained when the development and applications of that know-how involve large capital and labour, thereby it is costly compared to discharges into the environment. Furthermore, barriers for applications of such know-how about pollution prevention are encountered. For example, liabilities for pollution are obstructed when decision-makers do not care about the affected interests, obsolete technologies are protected when authorities enable the polluting activities with fail-soft regulations, and other market and policy barriers for pollution prevention. Regarding the signals of deficiencies, many opportunities for loss prevention can be perceived while not pursued because of barriers to innovations.

The growing know-how about pollution prevention generates higher value because increases deliveries of the demanded products and simultaneously reduce emissions. Costly developments and applications of that know-how for new products must be balanced by the sales of valuable products, as well as cost-reducing prevention of losses, and reduced risks of liability. If the sales, cost-savings, and reduced risks due to novel product designs exceed the costs, innovation-rents are gained; a rent is an income from assets. When the innovation-rents are allocated into labour and capital that add novel valuable qualities of products, additional incomes are generated from sales of these innovations while the losses are reduced. This allocation of the innovation-rents generates technical change called 'autonomous' because driven by the suppliers partially independent of the consumers' and policy demands. Such allocation of rents for novel qualities of products generates income growth along with emission reduction. A result is the decoupling of the growth of income from energy, materials, and emissions throughout the 1900s (Krausmann et al., 2009). Note that a part of the rents is accrued by shareholders as profits for consumption, and the rents are wasted when the designs fail. Besides the autonomous technical change, specific consumer and policy demands can invoke innovations for pollution prevention. In particular, the knowledge workers demand environmental qualities because they need spacious, clean, tranquil environments for the knowledge exchange and the policies pursue regulations for the common good in societies because the external effects can undermine productivity and welfare. Herewith, renewable energy is considered an innovation spur that is driven by autonomous and induced technical change.

Although technology is generally considered as a 'black box' technological change is much studied, referred to as 'innovation process'. The innovation process is usually comprehended as stepwise changes from an idea through technology development, sales of a new product, toward consumption. This process can address new technologies for production or new products; for example, new machines for the production of solar panels or new solar panels. Both can be attractive to customers because reduce their production costs or deliver beneficial qualities. In this process research and development (R&D) is supposed to deliver valuable novelties (inventions). The inventions are assumed to be used by entrepreneurs for sales of products and services (innovations), followed by growing purchases over time (diffusion). The supplying entrepreneurs are usually labelled as 'innovators', and the purchasers of innovations as 'adopters', though this distinction is unclear when innovations are applied within a firm; for example, gasification of bio-residues in a paper mill. The innovation processes involve costly know-how and equipment during the several years before any income can be generated. Meanwhile, that development can fail caused by deficient know-how and purchases can be low caused by changing demands or new competitors, but the innovation processes must be financed in advance, which involve risk-taking investments.

Innovations are pursued if the firms under competition expect to capture a market share that enables the generation of income and thereby provides an opportunity to make a profit. Innovations on market are considered a rivalry between firms using novel technologies and ones using technologies established in the past, called incumbents. A rivalry implies the choice of an alternative in decision-making. The conventional idea is that the rivalry between firms triggers substitutions for technologies that reduce the production costs of the innovator. Lower costs enable the generation of profits and provide benefits to customers through lower product prices. However, when competition is weak or distorted, the non-innovative behaviour of firms prevails (Kamien & Schwarz, 1982). In line with this viewpoint, firms that produce renewable energy are usually considered rivals of the firms that produce fossil fuels, though combinations of renewable energy and fossil fuels are also observed.

During the diffusion of innovations under competition, producers and consumers usually improve the performances of novelties, called ‘learning by doing’ when these improvements refer to production; and ‘learning by using’ when they refer to consumption. This technical learning reduces production costs and improves performance which enhances the productivity of economies (Rosenberg, 1982). Innovations are usually costly and imperfect when launched; and therefore, they are usually tuned to particular early adopters called a ‘market niche’, followed by the technical learning in time. As the innovation processes evolved during the 1900s toward nearly continuous innovative efforts in reduction of the production costs and increase of the value of the products, the innovators generated know-how about enhancing of the consumer demands, called ‘marketing’ (Mokyr, 2005). From the late 1900s on, many innovations were also generated by consumers that experienced deficiencies in their consumption. These consumers are tinkerers that invent and construct through trial and error, rather than undertaking the due processes of research, engineering, or design (von Hippel, 2005). Such ‘user innovations’ in energy consumption refer to a distributed energy system. The distributed energy systems are comprehended as small-scale, dispersed renewable energy production in local electricity and heat networks which are combined with storage, local energy services, and demand management. Such energy systems are also labelled ‘local energy systems’, ‘community energy’, and similar names, whereas the demand management refers to activities aiming at the change in energy consumption. Decision-makers in businesses and policies are usually preoccupied with larger scale of activities. Technologies in production usually reduce costs per unit output when they are applied on the large scale, referred to as economy of scale. The costs per unit output - called marginal costs or unit costs - decrease because investments in machines usually increase less than proportional to larger output. Meanwhile, the distributed energy systems that are small scale can be tuned to the specific customers’ demands which can increase the perceived qualities, improve performance in consumption at higher costs, or also reduce costs, called economies of scope. Although more scholarly and entrepreneurial attention is given to the economies of scale than to the economies of scope it is difficult to assess before applications of particular technologies which one can generate more social benefit or private profit per unit of energy.

Innovations disseminate through transactions between suppliers and purchasers in the value chains. Given competition between various suppliers, the innovations are pursued when suppliers can compound novel qualities into supplies if they expect to obtain a higher price or larger number of sales because new qualities are attractive to subsequent purchasers in the chain. This way, the innovators can cover the costs of innovation processes and make a profit, whilst purchasers in a value chain and ultimately final consumers can benefit from lower costs and better performances, given that price. Though high priced purchases are often associated with the snobbish shopping, so called 'Veblen goods', purchasers of novel designs and brands, so called early adopters, usually pursue a higher status at a price; be it, for example, solar panel, and Tesla car. In effect, transactions in value chains evolve toward an equilibrium price of a specific quality alongside additional qualities. Given customers income, those transactions enlarge the diversity of qualities because every subsequent supplier in a value chain tries to compound an additional quality with the aim accrue a larger market share. Meanwhile, subsequent customers pursue the qualities perceived as valuable, after deliberations about prices and deficiencies (Lancaster, 1966). The compounding of additional qualities generates specialisations, expressed as the diversification of activities in markets.

The deficiencies known to suppliers can be intangible to customers. However, the quality assessments are usually biased because the suppliers' interest is to show qualities assumed attractive to purchasers, and hide deficiencies that hinder sales. Meanwhile, the purchasers have the interest to present their preferences but cannot assess the deficiencies prior to purchases and uses. This asymmetric information persists unless the suppliers are liable for all deficiencies; however, their interest is to obstruct such liabilities (Akerlof, 1970). Moreover, deficiencies external to these transactions are preferably ignored. In particular, those interests are neglected that have limited capabilities to raise voices about deficiencies; for example, the voices of children, nature or the environment. Nevertheless, such deficiencies undermine welfare. For example, in the gasoline value chain, mined oil is refined, and transported to the fuel stations for consumption in engines; however, more than 90% of energy is lost along with emissions throughout the chain of these conversions. Even with state-of-the-art technologies, emissions caused by the losses during the conversions cause spills, smog, acidification, greenhouse gases, and other ill effects, while pollution is partly neutralised in nature due to the natural sinks in soils, oceans, and the atmosphere.

So far, several basic concepts in the economist comprehension of innovations are presented. Why and how innovations emerge is debated. This debate can be divided into three modes of thinking. A few key points in these ideas are summarised, as they help in assessing the mechanisms of change in energy production and consumption. The neoclassic argumentation, which is mainstream economics, is pre-occupied with prices based on transactions between producers and consumers; the evolutionary theorising is focused on knowledge and technology development, and the behavioural opinions address values in activities. These schools of thought are introduced, and linked to energy issues.

In the mainstream theory, selfish suppliers and customers aiming at least-cost preferences, called 'utilities', establish prices in transactions. In such transactions, scarce resources cause higher supply prices, followed by lower demands that invoke lower prices. If those interests are free and rational, this sequence of events evolves toward an equilibrium price wherein nobody can be made better off. The persisting demands, scarcities, and other market deficiencies trigger innovations that substitute incumbents, labelled as 'disruptions' (Christensen, 2000). These innovations generate new equilibria of supplies and demands at lower prices. This transactions mechanism called a 'market' is an axiom. Observed deviations are considered deficiencies that require corrections; for example, monopolies or cartels – which are the dominance of one or a few firms in a market – are assumed to undermine price negotiations, thereby opposed by authorities that are assumed to pursue the market mechanism. This market narrative is integrated in the rule of law. Legal are the market transactions, whereas non-market activities are considered as informal or illegal. In a result, many millions of people who cannot prove ownership based on the market transactions are dispossessed and persecuted. From this mainstream viewpoint, policies can influence market prices through regulations subsidies, and taxes. Therefore, the consumption of renewable energy can be enlarged through cheaper energy with subsidies and CO_2 emission can be reduced through a tax on emissions.

The global market of energy resources of about 6500 billion US Dollars (USD) a year grows at nearly 2% annual average, which is in line with the global income using data of the International Energy Agency (IEA). On the energy markets, supplied energy resources and energy products as electricity are traded with purchasers on auctions that generate prices of these commodities; meanwhile, energy resources as biomass in nature and combustion emissions are not priced. Scarce energy resources are assumed to increase energy prices. Higher prices of energy resources are expected to invoke technologies that generate lower-cost energy resources, called 'backstop technology'; for example, nuclear power was considered the backstop technology of coal, oil, and gas combustion for electricity production during the 1970s (Nordhaus, 1973). The backstop technologies are assumed to disseminate when they reach so-called price parity, which means similar costs to the lowest cost technology used in the energy networks. The price parity on electricity networks (grid) is usually assessed because the heat networks are underdeveloped. Due to commodity markets, the prices of energy resources are assumed to match the lowest cost technologies per energy unit in the life cycle, called the 'Levelized Cost Of Energy', abbreviated as LCOE (Usher, 2019). For comparison of various energy technologies, their LCOE is estimated with standardised interest rate, capacity factor, the lifetime of the installation, and other key variables in energy production.

Although it is aimed to provide objective cost estimates with upper and lower ranges, as well as the weighted averages unit costs – sometimes including the unit costs of winning tenders in energy projects – in practices, variations in the estimates are observed because conditions for the assessments differ (Timilsina, 2020). For this estimation and verification of results, the National Renewable Energy Laboratory (NREL) in the United States of America (USA) developed a calculation model of the LCOE (NREL, 2021a). However, the normative choices cannot be avoided, as the choice of applicants, and the life span of applications. Similar issues are faced in life cycle assessments (LCA), Energy Return on Investment (EROI) and other life cycle benchmarks. The production of renewable energy is promoted for being as costly as fossil fuels – per kWh 2–4 dollars-cent versus 1.5–3 dollars-cent, but it is pinpointed that the variable flows must also be distributed and stored, while such renewable energy systems are costly.

Evolutionary economics argues that the cumulative knowledge from the past generates innovations. The results of technological change determine the prices of market transactions, instead of the other way around as argued in the mainstream theory. When the specialisations and scales of production and sales during diffusion of innovations increase the costs of technologies per output decrease followed by the declining prices (Nelson, 1995). The main mechanism of the innovation process is that the cumulated knowledge generates options for the development of novelties, followed by the selection of options based on expectations about sales (Dosi, 1997). That mechanism is assumed to be generated by the negotiations and arrangements between policymakers, businesses and experts, labelled as the 'triple helix' model (Etzkowitz & Leydesdorff, 1995). Herewith, policies are assumed to invoke innovations when invest in public R&D and define regulations that enforce new technologies, so-called 'technology forcing', meaning policies that impose demands beyond available technologies. This idea is relabelled as the 'mission-oriented policy'; for example, the *Energiewende* in Germany aiming the ending with coal and nuclear power through larger renewable energy production along with reduction in dependency on energy imports (Mazzucato, 2018). Therefore, policies enforce demands for higher performances combined with the support of innovations that attain these policy demands.

In energy, a higher energy efficiency is considered a key factor. The energy efficiency is considered a higher energy service in heat or power per resource mass or percentage energy service to energy resource; the energy service is defined as a desired end-use or state (Fell, 2017). Given that every conversion of the energy resources for energy services causes some losses, additional conversions in the value chain cumulate energy losses and costs, whilst technical change for energy efficiency reduce the losses. This change is referred to as the 'effect-increasing technical change'. An example is global energy efficiency in production increased from about 10% in the early 1800s to 34% in the early 2000s despite more conversions in the value chains. However, the global energy efficiency in consumption remains below 10% with some exceptions. An example is lighting whose energy efficiency improved from less than 20% in the light bulbs to more than 90% in the LED measured by light performance (lumen) to electricity consumption (Grübler et al., 2005). As result of higher energy efficiency and larger scale of LED light production and sales, the cost of lighting per unit of service decreased called the 'cost-reducing technical change'. Therefore, the development of knowledge for energy-efficient technologies is advocated and considered more effective as incentives for innovations compared to putting higher taxes on energy resources which generate high prices of energy consumption (Jacobsson & Bergkek, 2004). From this perspective, deficient markets, poor infrastructure, imperfect institutions, poor communication, and other barriers for the dissemination of renewable energy are pinpointed, and the policies are assumed to resolve the problems (Negro et al., 2012).

Higher energy efficiency saves costs. Less costly energy enables larger expenditures on energy consumption which increases welfare – if valuable services are consumed. This way, higher energy efficiency also invokes additional extraction of energy resources with associated pollutions, called the 'rebound effect'. The welfare growth due to higher energy efficiency is often indicated by lower energy consumption per income, called 'energy-intensity'. Estimates show that the global energy-intensity measured by the global energy consumption per GDP has grown by 1%–3% a year during the second half of 1900s; it means less energy is needed for income generation (WEC, 2010). In this book, its reciprocal is used which is the income per energy unit, called 'energy performance'. This phrase is used for indication whether more income is generated per energy unit. Similarly, the income per CO_2 is called 'carbon performance'. Higher energy performance indicates a higher contribution of energy services to income, and higher carbon performance implies lower damages caused by energy services.

The behavioural approach addresses the values, and behaviour of individuals, groups, and organisations. In this train of thought, the market and policy imperfections are assumed persistent because they are subjected to deficient assessments of a fundamentally uncertain future. Innovations are driven by the entrepreneurial capabilities to resolve partially these imperfections and generate profit; scanning of profitable opportunities based on the deficiencies is considered key entrepreneurial capabilities (Kirzner, 1997). Therefore, finding opportunities tuned to particular demands drives innovations. Innovations disseminate when they are adopted within a small core group of consumers that scan novelties to match their values, given prices, followed by dissemination across various consumer groups along with decreasing prices because competing suppliers introduce adaptations that reduce costs (Rogers, 1983). The values refer to the functional and ethical qualities of novelties. The functional qualities address mechanics, heat, density, and other technical properties that enable uses. For example, the value of gasoline is determined by hydrocarbons, sulphur, octane, and about a dozen other qualities which are compounded into a few indicators at the pump. The functional qualities are usually decisive in businesses. For the final consumers, ethical qualities are also essential. For example, convenient, autonomous, natural, and other ethical qualities express attitudes, and social associations in consumption called 'lifestyle'. High ethical qualities generate price mark-ups for products, goodwill for investments, awards for performances, and other prizes, thereby contribute to customer satisfaction (Kotler and Zaltman, 1971). The ethical qualities gain importance as economies evolve from material-intensive industries toward knowledge-based services (Stahel, 2010). In this argumentation, the policy is not an autonomous factor but driven by interests in society. Innovations emerge when a sense of urgency for changes is generated with regard to imperfections in markets and policies and disseminate due to individuals and groups that are capable of attracting supportive adopters in markets and policies, whereas the political decision foster or impede those innovators (Krozer & Nentjes, 2006).

Conversions of energy resources deliver services that generate functional and ethical qualities for consumption. Examples of valuable qualities to consumers are heat, power with respect to functional qualities, as well as cleanliness, and convenience with regard to ethical qualities. From the 1990s on, energy services based on modern renewable energy absorbed larger investments, than traditional renewable energy, and grew faster than cheaper fossil fuels as they generated benefits as perceived by consumers. Major beneficial qualities of modern renewable energy are low pollution compared to fossil fuels which enable the decarbonisation of economies, and a value addition by its widespread availability because water, wind, and solar resources are found virtually all over the Earth. That value addition refers to, for example, the consumers' sense of autonomy that invokes energy production by consumers called 'prosumption' (Toffler, 1980); a sense of belonging to communities that triggers distributed energy systems (Ornetzeder & Rohracher, 2006); as well as a lifestyle that demonstrates commitments to the mitigation of climate change (Plumer, 2016). Value added by energy services can be reached through better tuning to individual and societal demands; among others, a wider access to energy and CO_2 emission reduction. Such value added justifies growing interests in the distributed energy systems.

Those theories contribute to a better understanding of the mechanisms of change in energy production and consumption. Rather than testing the validity of these theories based on empirical observation, the energy prices, energy efficiency, and valuations are used for the interpretation of the statistical data.

1.3 Data and Method

As mentioned, the mechanisms of change are largely based on interpretations of the statistical data. Therefore, several authoritative, public statistical databases are used. Those frequently used are mentioned below:

General:

- World Bank (1960–2017) data – WB https://data.worldbank.org/
- Eurostat – https://ec.europa.eu/eurostat/home? (1990–2017)

Economic:

- Historic economic data based in Maddison (2007) and substantiated in the Maddison project database, version 2018 (www.ggdc.net/maddison).
- USA Census on the business data – https://www.census.gov/programs-surveys/susb/data/datasets.html

Combined Energy and Economic Data:

- Historic energy and economic data in the database 'Our World in Data' (OWD) – (https://ourworldindata.org/)
- PBL Netherlands Environmental Assessment Agency, Historic Database on Global Environment (HYDE) – https://themasites.pbl.nl/tridion/en/themasites/hyde/index.html

Energy:

- The International Energy Agency (IEA) for most energy data 1990–2015 – https://www.iea.org/classicstats/statisticssearch/ until 2015; later https://www.iea.org/data-and-statistics/data-tables?country=WORLD
- BP-energy (1965–2019) – https://www.bp.com/en/global/corporate/energy-economics/statistical-review-of-world-energy.html
- Ritchie, H., & Roser, M. (2020). *Energy, Published online at OurWorldInData.org*. Retrieved from: https://ourworldindata.org/energy [Online Resource]. Accessed 21 Mar 2020.
- Bloomberg, Frankfurt Business School and UNEP (2004–2018) on investments in renewable energy https://about.bnef.com/energy-transition-investment/
- International Renewable Energy Agency (IRENA) on installed capacities, production, investments and unit costs (2000–2019) https://www.irena.org/publications/2020/Jul/Renewable-energy-statistics-2020
- Energy Information Administration (EIA) 1990–2017 – https://www.eia.gov/
- United Nations, World Energy Supplies in Selected Years, 1929–1950, Statistical Papers no 1. https://unstats.un.org/unsd/energy/yearbook/Series_J_No_1.World_Energy_Supplies-1929-1950.pdf
- Database on energy production, largely based on Etemad and Luciani (1991), – https://theshiftdataportal.org/

In addition, several authoritative assessments are employed; for example, Lazard on the costs and effects of renewable energy, Renewable Energy Network (REN21) on policymaking, and Statista on energy production in the USA.

Interpretations of these data were used for the assessments of the mechanisms of change in the past whereas models for scenarios about the future are avoided. Furthermore, the interpretations of the data are based on simple mathematical formulas. The calculations are explained in words. For the convenience of the readers, nearly all formulas used in the book are shown in Appendix 1. All calculations are transparent and made in Excel. They can be delivered by the author of this book solely through a personal request by mail (krozer@xs4all.nl).

Mainly income, energy and CO_2 are assessed. The income refers to Gross Domestic Product (GDP), which is by definition equal to the National Income. In many assessments, the GDP is corrected for inflation based on a typical consumer purchase basket, referred to as 'Gross Domestic Product in Purchasing Price Parity' (GDP-PPP). Energy and electricity are shown in power units (kWh) as they are widely used in the billing, contrary to the conventional energy units in joules and calories that are rarely used for bills. Energy resources and CO_2 emissions are usually in tonnes.

The assessments are focused on the world as a whole, and the largest countries measured by population. The countries that exceeded 100 million inhabitants in 2017 are selected, with the European Union of 28 member countries (EU28) considered as one country; this includes the United Kingdom that left the EU in 2020. A few assessments also cover countries within the EU. The countries in focus in the ascending order of GDP-PPP per capita in 2015 are: the United States of America (USA), Japan, the European Union (EU), which are considered high-income countries (above USD-PPP of 30,000 per capita in 2015); followed by the Russian Federation (Russia), Mexico, Brazil, Republic of China (China) and Indonesia, which are called mid-income countries (between USD-PPP 10000 and 30,000 per capita); and Philippines, India, Nigeria, Pakistan, Bangladesh and Ethiopia, which are considered as low-income countries (below USD-PPP 10000 per capita). Based on the IEA data for 2015, these countries together covered about 70% of the global population, 74% of the global GDP-PPP, 65% of global energy production, 72% of energy consumption, 76% of electricity consumption, 88% of coal supplies, 63% of oil, 64% of gas, 83% of nuclear energy, 71% of biofuels, 66% of hydropower, 84% of modern renewable energy and 73% of the global CO_2 emissions. It means that these countries determined the global changes.

The income and energy consumption varied across countries in 2015. Table 1.1 shows a few indicators in total and per capita in 2015. It is based on World Bank data. The indicators are population, GDP-PPP, energy consumption, electricity consumption, CO_2 with deviations between IEA and World Bank data and percentage renewable energy in energy consumption. In 2015, the spread between the highest and lowest data across those countries was huge. While the GDP per capita in Ethiopia was about 35 times lower than in the USA, the lowest energy consumption per capita in Ethiopia compared to the USA was 14 times smaller; electricity consumption was 150 times smaller; CO_2 emissions were a few hundred times smaller, and the share of renewable energy in the total energy consumption was

Table 1.1 Indicator data for the selected countries

	Population	GDP-PPP	Energy	Electricity	CO_2 (IEA/WB)	Renewable energy
Data for 2015	Total	USD_{2010}	kWh/capita	kWh/capita	million tonnes	% Energy consumed
World	7355	15,694	22,579	3147	32,294 (89%)	18%
USA	321	56,469	81,330	12,975	4998 (93%)	9%
Japan	127	40,607	39,137	7651	1142 (97%)	6%
EU	510	38,447	34,209	5752	3201 (105%)	17%
Russia	144	24,692	55,763	6666	1469 (90%)	3%
Mexico	126	16,983	16,922	2101	442 (94%)	10%
Brazil	206	15,617	17,617	2651	451 (81%)	41%
China	1371	14,450	26,213	4081	9041 (88%)	13%
Indonesia	258	11,040	10,495	851	442 (101%)	37%
Philippines	102	7320	5773	713	104 (92%)	28%
India	1309	6127	7749	846	2066 (84%)	35%
Nigeria	181	6038	8664	146	64 (68%)	88%
Pakistan	189	5000	5573	460	146 (87%)	46%
Bangladesh	161	3336	2658	328	71 (92%)	37%
Ethiopia	100	1633	5787	75	10 (81%)	91%

more than 10 times larger. The highest energy consumption, and electricity consumption per capita was in high-income countries, as well as in Russia and China which are mid-income countries; whilst China has high share of renewable energy, Russia has low. By far the largest CO_2 emissions is in China and the USA, which together covered 43% of the global emissions. Low-income countries used less energy, caused less CO_2 emissions and had higher shares of renewable energy in the energy mix in comparison with high-income countries and mid-income ones. Herewith, the IEA data on global CO_2 are about 11% below the World Bank data; it is 5% higher in CO_2 per capita in the EU, but 32% lower in Nigeria. Deficient data undermine coherence in policies aiming at the mitigation of climate change because the international policies are based on the voluntary reporting about CO_2 emission reduction by each country.

All these countries' variables change over time. Moreover, some changes can evolve in different directions during a long period of time, called the diverging trend, whilst other changes can evolve in a similar direction, called the converging trend. These trends are assessed based on the annual average growth rates of the countries' changes, measured by the standard deviation. An increasing standard deviation during minimum of ten years indicates a diverging trend, and a decreasing one shows a converging trend. For example, the diverging trend in CO_2 emissions means the increasing differences in the growth rates across countries whilst the converging trend in energy consumption implies the decreasing differences in their growth rates. Furthermore, a higher standard deviation than the average indicates a high spread of the results, which implies a disputable conclusion.

A major issue is whether the renewable energy data are reliable because measurements of energy production and consumption are imperfect. The statistical data on energy resources and consumption is reasonably available and reliable insofar it relates to the sales of heat and electricity on markets and is subject to taxes or subsidies because these are usually observed and published in the countries' statistics. This data is systematically published from the late-1800s on; Romania, being part of the Austro-Hungarian Empire at that time, is the first country with the oil statistics from 1857 on. So, the energy data on earlier periods are estimates incidentally done by various authors using their own methods. Herewith, intervals of 10 years are used for assessments of changes during two centuries, and interpretations based on this data address trends throughout decades rather than particular years. Furthermore, not all countries compile reliable data which causes data imperfections. Therefore, databases are compared as much as possible because even basic data in those databases differ.

Figure 1.1 shows data on renewable energy for electricity generation from 1990 to 2015 in five-year intervals, which is derived from five authoritative databases. IEA data in the energy balances are based on data delivered by governments; OWD database states that this data combines estimates of Smil (2017) with BP data; SHIFT Portal states that it is largely based on the estimates of Etemad and Luciani; BP and IRENA databases are based on their expertise. All data are shown in Terawatt hour (TWh); fuels in Petajoule (PJ) is equal to 0.278 TWh (PJ = 0.278 TWh). Those 25 years between 1990 and 2015 can be divided into periods of high and low oil

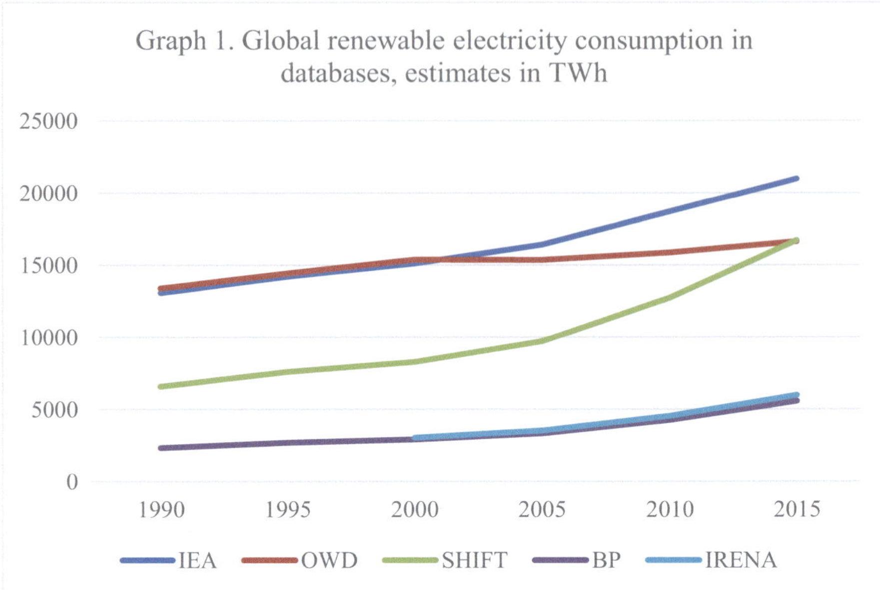

Fig. 1.1 Global renewable electricity consumption in databases, estimates in TWh

prices. Note that oil prices are considered as trend-setting for the prices of other energy resources. The oil prices are usually estimated in USD per barrel oil equivalent (b.o.e. or bbl.); 1 b.o.e. is equal to about 136 kg oil, or 7.4 b.o.e. in tonne oil. Based on the prices data of BP, the oil prices fluctuated annually measured in real dollars in 2019. They increased from USD_{2019} 20–52 per b.o.e. during 1990–2004 to USD_{2019} 56–126 per b.o.e. during 2005–2015. Inter alia, these periods are somewhat arbitrarily defined as low fuel prices emerged in the mid-1985 after high prices during 1974–1985, and the prices started to increase in 2004 and dropped mid-2015 to the level of 1990s. Furthermore, the prices of gas increased in line with the oil prices, but the prices of coal and nuclear resources hardly increased; the prices of fossil fuels are discussed in more detail in the Sect. 5.2.

The largest consumption of renewable energy is found in the IEA database, whereas the OWD database shows similar consumption until 2000, though slower growth onward. Data in the IEA and OWD databases are similar during the 1990s, possibly based on the same data sources. Thereafter, the IEA data shows faster growth of renewable energy, which is plausible because renewable energy was attractive relative to fossil fuels during high fuel prices. SHIFT database shows lower production in 1990 and faster growth after 2005; and by 2015, OWD and SHIFT show similar production. BP and IRENA production data are lower, similar to each other. The BP and IRENA data are similar, but not exactly the same, and both show slower growth than the IEA database. Different scales can be explained by definitions because the IEA and OWD based on IEA include biomass and household waste. Those differences in the growth of renewable energy are caused by assumptions in calculations.

The highest data on the consumption of renewable energy was 6 times higher than the lowest one 6 during the 1990s and 4 times during 2010s. The average annual growth during that period varied from 1.6% to 3.8% in most databases, whilst it was implausibly low 0.9% in the OWD database. Periodic fluctuations in the growth rates are larger. There are also differences in data across the energy resources. For example, the production of wind and solar energy together in the SHIFT database is three times larger than in the other databases. An explanation can be that the SHIFT database includes solar heaters and electricity, while other databases cover electricity only. Furthermore, the scale of electricity generation from bioenergy in 1990 is twice larger in the SHIFT database compared to the IEA database which is twice larger than in the BP database. Differences in the data are partly caused by the variety of assessment methods. These differences are worrisome because decisions and policies are biased by such deficiencies; it begs one to assess just how reliable projections about the future are when the estimates about the recent past vary that much. Insofar as is possible, databases are compared before concluding about the changes and trends without any verdict about the reliability of particular databases.

Besides those variations, there are also different definitions in the databases; they partly explain the differences in quantities. The IEA database carries the broadest definition of renewable energy. This definition encompasses: municipal waste, industrial waste, primary solid biofuels, biogas, liquid biofuels, geothermal, solar thermal, hydropower, solar PV and tide, wave ocean put together. The OWD does not specify these resources; nevertheless, its definition is possibly similar to the IEA. The SHIFT database is more specific about liquid biofuels as it specifies bioethanol and biodiesel. The IRENA database specifies wind power onshore and offshore, as well as solar PV and Concentrated Solar Power. The BP database has the narrowest definition that covers hydropower, geothermal energy combined with bioenergy, wind, and solar energy. Herewith, renewable energy is considered in the broad sense of the IEA, though particular attention is given to modern renewable energy that is defined as geothermal energy, wind power, solar power, and biofuels which are used for heat and electricity.

Based on the IEA data, the composition of energy resources in the global consumption of heat and power in 2015 is shown in Table 1.2, excluding calories for food and feed. The shares and growth rates are estimated for the period of low prices of fossil fuels from 1990 to 2005 and high prices of fossil fuels 2005–2015; these growth rates are annual averages of the five-year intervals. All data are shown in PWh (= 10^{15} kWh) using the IEA energy balances in million ton oil equivalents (Mtoe) that are converted in power units, assuming one Mtoe is about 0.01163 PWh.

Most energy resources consumed in 2015 were fossil fuels. Whilst coal, oil, and gas shares in the total energy consumption were 22–32%, the share of nuclear energy consumption was only 4% that year. While all fossil fuel resources grew a few percent per year during low fuel prices, the growth of oil declined during high fuel prices and the nuclear energy consumption decreased in total. The latter cannot be linked to the fuel prices because nuclear power does not depend on the prices of other fossil fuels, but high costs and unresolved safety problems are the main reason for closure of the nuclear power plants in several countries and delays in the constructions and deployment of new ones. Renewable energy resources covered

Table 1.2 Global resources for energy consumption in PWh, excluding calory in food and feed

IEA data in PWh	Consumption in 2015	Shares in all energy resources	Annual average growth	
			1990–2005	2005–2015
Coal	45	28%	2.0%	2.5%
All liquids	50	32%	1.4%	0.8%
Gas	34	22%	2.4%	2.2%
Nuclear	8	5%	2.3%	−0.7%
Bioenergy	15	10%	1.5%	1.6%
Hydropower	4	2%	2.1%	2.9%
Geothermal, solar wind	2	1%	4.4%	11.1%
Fossil fuels	137	86%	1.9%	1.6%
Renewable	22	14%	1.7%	2.5%
Total	159	100%	1.8%	1.7%

Table 1.3 Modern renewable energy TWh electricity and heat

IEA data in TWh; PWh = 1000 TWh	Consumption in 2015	Shares in renewable energy	Annual average growth	
			1990–2005	2005–2015
Liquid biofuels	9	0.04%	89%	13%
Geothermal	90	0.4%	3%	3%
Solar thermal	10	0.05%	0%	32%
Solar PV	250	1.2%	28%	52%
Wind	834	4%	25%	23%
Marine[a]	1	0.00%	0%	7%
Modern renewable energy	1194	6%	10%	21%
All renewables processed	21,614	100%	3%	5%

[a]Tide, wave, and ocean

about 14% of all energy consumption. The largest resource is bioenergy, including consumption of agricultural and forest residues for heating which cover the largest part of bioenergy as the generation of power from biomass and waste is only about 10% of that total. Another large renewable resource is hydropower. Bioenergy and hydropower constitute the traditional renewable energy resources which covers 12–13% of all energy consumption. Nearly 2% of energy consumption is covered by modern renewable energy which includes geothermal energy, wind energy, solar energy, biofuels, and a few other resources at negligible quantities at this moment. While fossil fuels and traditional renewable energy grew a few percent per year during low and high fuel prices, the modern one grew twice faster during low fuel prices and about 11% during high prices.

Large differences in the scale and growth rates can be observed within modern renewable energy. Table 1.3 shows these resources.

Modern renewable energy covered only about 6% of all renewable energy resources in 2015 while biomass and hydropower covered the remaining 94% of all. Liquid biofuels, geothermal, and solar thermal were mainly used for heating whereas, solar photovoltaic (PV), wind, and marine resources were used for electricity production. Among six main resources in modern renewable energy, only wind and solar energy had a significant scale, which was 1.2% and 4% of all renewable energy, whereas, four other resources were small scale in 2015. Wind energy grew by nearly 25% throughout 25 years, solar photovoltaics (PV) for electricity grew even faster as the growth rates were 28% during 15 years of low fuel prices and 52% during 10 years of high fuel prices. Heating based on biofuels grew even faster during low fuel prices and solar thermal expanded during high fuel prices, but their scale is limited to the market niches. Those growth rates are performed during 10 years or longer. Such high growth rates determine the speed of energy transition because 25% annual average growth implies doubling of the scale after 3 years and 1000 times larger scale after 30 years, whereas, 35% doubles the scale within 2.5 years and reaches 1000 times larger scale after 23 years. Therefore, the scale of an energy resource at a particular moment in time is not as relevant as the continuously high growth. In this book, attention is mainly given to the growth of renewable energy during long periods. As the growth rates somewhat differ across those databases and per period, the databases and the years of estimates are noted when changes are assessed.

For the estimates, the conventional engineering conversion are much used. The engineering databases on conversions are used as the energy content of resources varies. The energy content is measured as energy density, meaning energy that can be generated from a unit of mass in standardised conditions. The energy densities of commonly used fuels are shown in Table 1.4. The energy density of agricultural biomass is lower than dry wood that is somewhat lower than coal; and about half lower than oil. By far, the most energy dense resources are: hydrogen because it is nearly three times denser than oil, and uranium oxide for nuclear power. These energy densities are used despite minor differences between engineering databases. The energy content of resources is often measured in the non-metric scales. Non-metric scales cannot be avoided and they are often converted into the metric one. The most popular non-metric scales are mentioned below with equivalents similar to the metric system. These are: 1.4 tonne of coal is similar to ($\approx$) 1 tonne of oil equivalent (t.o.e.) $\approx$ 5.8 GJ, $\approx$ 11,630 kWh, $\approx$ 7.4 barrel oil equivalents (b.o.e. or bbl.) $\approx$ 1395 Nm^3 natural gas, $\approx$ 1 tonne of gas (LPG). 1 Nm^3 gas is similar to 35.3 cubic feet (cf) $\approx$ 1.83 kg, $\approx$ 0.0009 t.o.e. $\approx$ 10.5 kWh; in the USA, 1 million BTU gas (MMBTU) $\approx$ 293 kWh. Herewith, non-metric system is usually converted into kWh. Food calories are rarely used in the estimates, except for heating and the labour of people and animals; 1000 calories (cal) or 1 kilocalorie (kcal) $\approx$ 1.16 MJ $\approx$ 0.32 kWh. Many engineering converters can be found on the web, among others: https://www.unitconverters.net/energy-converter.html

Basic engineering concepts are used for economic assessments. For a time, it is 3600 seconds in 1 hour (h); 24 hours a day; 365 days a year (y) which equals 8760 hours a year of full capacity applications. Most energy measurements refer to

Table 1.4 Energy density and carbon density of several energy resources based on the Engineering toolbox, 2020, https://www.engineeringtoolbox.com/

Resources in MJ per kg resource (kWh using MJ ≈ 0.278 kWh)	
Peat, Grasses	17 (4.7)
Wood Dry	20 (5.6)
Ethanol	27 (7.5)
Biodiesel	44 (12.2)
Brown Coal (lignite)[a]	15 (4.2)
Black Coal	26 (7.2)
Crude oil, gasoline, diesel	43 (12.0)
Natural gas, LNG[b]	49 (13.6)
LPG[b]	47 (13.1)
Hydrogen	120 (33.3)
Uranium Oxide	470,000 (130660)
Carbon density, kg CO_2 per kWh (kg CO_2 per kg resource, IEA, 2017)	
Coal	0.32 (3.80)
Oil	0.23 (2.53)
Natural gas	0.19 (2.16)

[a]https://neutrium.net/properties/specific-energy-and-energy-density-of-fuels/
[b]https://web.archive.org/web/20100825042309/http:/www.ior.com.au/ecflist.html

Joule (J), it is the work needed to move 1 kilogram (kg) over a distance of 1 metre (m). While moving 1 kg on 1 kilometre (km) equivalent of 1000 kg (tonne) on 1 m is 1000 Joules or kilojoule, moving 1 tonne over 1 km is million Joules or megajoules (MJ); and so forth in multiplications of those. Units of power in kWh or multiplication of it are mainly used because many people are accustomed to them in electricity bills. One kilowatt-hour (Wh) is equivalent to moving 1 tonne kg per hour (kWh) or 1 tonne per 3600 seconds. Work and power are linked. Converting MJ into kWh means power needed to move 1000 kg over 1 km, which is calculated by 1000*1/3600, or 1 MJ = 0.278 kWh. For illustration, 1 kWh is used by a lamp of 20 W that shines 500 hours a year which is the typical LED lamp during 1 year in a middle-class living room, equivalent of elevating a basket of water 10 km high. Installations for energy production are often measured as the capacity of power in kilowatt (kW), or multiplication of it; 1 kW is equivalent to 1.4 horse power in engines. For the conversion of kW to kWh, the capacity is multiplied by 8760 h in a year, times percentage capacity utilisation in a year, times energy efficiency in deliveries. For example, 1 kW engine with 80% of the capacity utilisation and 50% energy efficiency delivers $1 \times 8760 \times 0.8 \times 0.5 = 3504$ kWh.

These conversions are not repeated in the text, while reading is possible without checking those formulas. All monetary terms are in market prices, excluding the valuations of beneficial and damaging qualities which are not used because contentious.

1.4 Chapters and Sections

After this introduction, the mechanisms of change are assessed in five chapters, ending with a seventh chapter that offers conclusions.

Energy resources and pollution are discussed in Chap. 2. The question is answered whether there are sufficient energy resources for larger energy consumption and lower pollution. After a brief introduction to this chapter, the flows of renewable energy are assessed in Sect. 2.2; the stocks of fossil fuels in Sect. 2.3; major changes in energy consumption are estimated in Sect. 2.4; changes in pollution in Sect. 2.5; and conclusions about resources and pollution are presented in Sect. 2.6.

Economic drivers of the shifts from traditional renewable energy to fossil fuels and then, to modern renewable energy are addressed in Chap. 3. The question is answered what drivers changed energy consumption in the past centuries. After a brief introduction, the changes in population, income, and energy consumption are indicated in Sect. 3.2 of this chapter. Those observations are interpreted for energy consumption from the perspectives of the neo-classic economics focused on prices, evolutionary economics preoccupied with energy efficiency and behavioural one about values in Sects. 3.3, 3.4 and 3.5 of this chapter, respectively. Conclusions are drawn in Sect. 3.6.

The innovation processes in renewable energy is addressed in Chapter 4. The question is discussed: what are the possibilities for innovations in renewable energy? After the introduction, Sect. 4.2 of this chapter addresses the lead-time in the development and dissemination of basic energy technologies during the two last centuries. Section 4.3 is about the business interests in renewable energy when compared to all sectors. Section 4.4 shows the chances for successful innovations in renewable energy, followed by Sect. 4.5 about the financing of these innovations. Conclusions are outlined in Sect. 4.6.

The emergence and expansion of renewable energy are addressed in Chap. 5, tackling the question-why did renewable energy grow? After the introduction, the energy prices, energy subsidies, and policy support for energy are discussed in Sects. 5.2, 5.3, and 5.4 of this chapter, respectively. Thereafter, the front-running stakeholders of modern renewable energy are introduced in Sect. 5.5; followed by an assessment of the start-ups and employment in Sect. 5.6; the decision-making systems, acceptance, and beneficial change Sects. 5.7, 5.8, and 5.9, respectively. Conclusions are drawn in Sect. 5.10 of this chapter.

Global consumption of renewable energy during the past decades is assessed in Chap. 6. The question is discussed whether energy consumption across countries became less carbon-intensive and more valuable. After a brief introduction to this chapter, the decarbonisation of economies in the past is assessed in Sect. 6.2; and the decarbonisation options with hydrogen in Sect. 6.3. Then, distributed energy services are assessed in Sect. 6.4. The value added by energy services in Sect. 6.5, followed by the valorisation of energy services across countries and its impact on global income, energy access, and CO_2 is considered in in Sect. 6.6. Conclusions are drawn in Sect. 6.7.

Chapter 7 draws conclusions about the mechanisms of change for accessible renewable energy.

Chapter 2
Energy Resources and Pollution

Can energy resources meet the growing energy consumption of present and future generations without undermining environmental qualities? This question is addressed using authoritative assessments on renewable energy and fossil fuels, and statistical data on energy consumption, and CO_2 emissions. While available resources of renewable energy enable a few thousand times larger than present annual energy consumption, well-exploitable resources of fossil fuels can be exhausted within several generations, with the exception of coal. The problem is that fossil fuels have complemented rather than substituted renewable energy in the past whilst modern renewable energy complements rather than substitutes fossil fuels at present. The growing fossil fuels also cause growing CO_2 emissions. Although the growth saturates, far-reaching emission reduction is needed. Sufficient emission reduction can be attained if 4–5% annual growth of renewable energy over the next decades substitutes fossil fuels.

2.1 Introduction

The pressing issues is whether sufficient energy resources are available for the accessible energy consumption regarding the growing global demands and whether these demands can be satisfied along with the mitigation of climate change. Without speculating about the energy demands in the far future – it is assumed to increase several times during the next decades – a larger scale of energy consumption must be met by energy resources that can be exploited without waiting for technologies whose viability and impacts are unknown; for instance, generating energy in space. Therefore, the present energy consumption is compared to the potentially available flows of renewable energy and stocks of fossil fuels. This comparison is followed by the explanation of changes in energy consumption throughout the two last centuries and the estimation of possibilities for sufficient reduction of CO_2 emissions.

Y. Krozer, *Economics of Renewable Energy*,
https://doi.org/10.1007/978-3-030-90804-1_2

With a given global energy consumption of about 572 EJ in 2015, equivalent to 159 PWh (10^{12} kWh), around 23,000 kWh per person was consumed. That consumption was divided amongst: 30% industries; 29% transport; 21% residential; 8% commercial and institutional; 1% agriculture and forestry and 10% all others. The consumption is skewed across countries because the USA and China consumed 38% of that total, and together with the EU this amounted to 50% of global energy consumption. While a person consumes on average 81,000 kWh in the USA, that average is 30 times lower in low-income countries such as Bangladesh, but 1600 times higher compared to a poor one that consumes only 50 kWh per person, labelled as 'energy poverty' by the International Energy Agency. The manifold larger scale of energy consumption can be expected as mid-income and low-income households in low-income countries generate a higher income because relatively more income is spent on this basic good. No doubt that they need it. Furthermore, energy consumption grows fast in mid-income countries as their private incomes grow fast because people demand more units of energy per unit of income. Meanwhile, energy saving in high-income countries evolves slowly.

Since all renewable energy on Earth is obtained on account of the irradiation of solar energy from space, that inflow must surpass the yearly global energy consumption of nature and people, now and in the far future. Nuclear fusion, moon reflection, and other geo-engineering technologies that aim to mimic or steer solar inflows are sought after as the 'Gulden Fleece' for energy production; however, all of those fall far from proof of concept, let alone being able to capture a large share in the global energy production within a few decades. Earth heat is also a source of energy. However, the Earth's heat from layers deeper than a few thousand metres, experienced as volcanic eruptions on surface, is not possible to exploit in the decades to come because technologies aiming to control that heat for safe exploitation are unreachable, as yet. For the time being – presumably, during a longer period of time than this century- global energy consumption of renewable energy depends on the use of solar inflows and outflows on the Earth. This is not a major problem because the net energy inflow is huge compared to all energy consumption by nature and people as shown in Fig. 2.1.

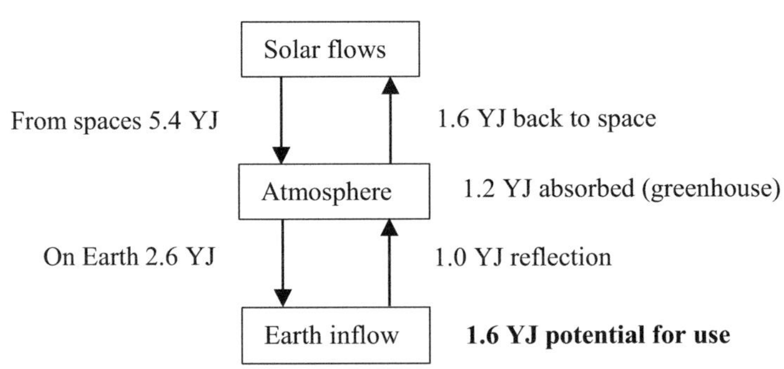

Fig. 2.1 Simple scheme of the solar flows on Earth in million EJ (YJ), using Rogner (2000)

An assessment of the solar flows on Earth (Rogner, 2000) shows that annually about 5.4 million Exajoule (YJ) – one EJ is 10^{12} MJ or 0.278 × 10^{12} kWh – enters the outer layer of the atmosphere, but 29% of that is reflected back to space by the outer layer of the atmosphere. About 1.2 YJ a year is absorbed in the atmosphere as vapour, gases, and particles in the air, which creates a greenhouse. The greenhouse effect of the atmosphere enables a liveable climate on most parts of the Earth. Without it, all solar irradiation would be reflected from the Earth back to space, while the average temperature on Earth would remain below −17 °C. The remaining 2.6 YJ a year flows to the Earth's surface. This energy inflow is absorbed between the Earth's surface and atmosphere. The hydrosphere of oceans and rivers absorbs about 2.1 YJ of that. Hence, water in solid form, as liquid, and vapor accommodates huge energy on Earth; it is not surprising that the Earth is often called the 'Blue Planet'. The lithosphere, which is a 2–100 kilometres thick outer layer of the Earth, absorbs about 0.5 YJ. The biosphere, which covers all plants and animals including humans, absorbs only about 0.004 YJ. More than 1 YJ of the inflow on Earth is reflected back into space. On balance, about 1.6 YJ per year of solar energy, which is about 445 × 10^{15} kWh, is available for consumption in nature, including for humans; this is about one-third of the total solar inflow that hits the atmosphere. The available solar inflow is sufficient for a global consumption that is about 2800 times larger than what it is at present.

That available solar inflow is apparently reachable, which means that it is possible to capture. This is referred to as the irradiation of solar energy. A large part of that reachable inflow can be captured on land, which seems to be easier than on the open seas where the transportation distances, storms, and other factors impede the exploitation of solar energy. It can be assumed that 600 Watt solar irradiation per m^2 land area hits Earth, which is after corrections for the Earth's orbit around the Sun, angle of solar inflow, locations, and other factors (Ghosh & Prelas, 2011, pages 80–86). Estimating on the assumption that 10% of irradiation can be captured with present devices during the average 2300 sunny hours a year, about 0.025% of 510 million km^2 land surface needs to be covered with the available solar energy technology to meet our present energy consumption. A larger area would be needed to cope with the growing energy consumption, despite improvements in the conversions of solar irradiation into solar power.

This estimate implies that the irradiation of solar energy on land is sufficient for a manifold larger scale of energy consumption than in 2015, but there are limitations for energy consumption in the far future.

2.2 Renewable Energy Resources

Based on that solar inflow, various resources of renewable energy are available. They can only partly be applied in addition to each other because the technologies for using them can be rival; for example, solar panels do not allow large-scale biomass production. Table 2.1 shows vertically renewable energy divided into the

Table 2.1 Annual use, technological potential and flow of renewable energy based on Global Energy Assessment

In EJ = 10^{18} J; use means consumption	Consumption 2005	Technological potential		Annual energy inflow	Indicators: min. is minimum, max is maximum	
		Minima	Maxima		Potential min. to use	Flow max. to potential
Biomass[a]	46.3	160	270	2200	3	8
Geotherm	2.3	810	1545	1500	352	1
Hydro	11.7	50	60	200	4	3
Solar	0.5	62,000	280,000	3,900,000	124,000	14
Wind	1.3	1250	2250	110,000	962	49
Ocean	*0.2*[b]	3240	10,500	1,000,000	16,200	95
Total	62[c]	67,510	294,625	5,013,900	1084	17

[a]Includes waste
[b]Resch et al. (2008)
[c]All energy consumption in 2005 was 484 EJ (IEA, 2015)

resources of biomass, geothermal, hydro, solar, wind and ocean, and horizontally the availability of those resources based on the Global Energy Assessment (Nakicenovic, 2012). Their availability refers to energy consumption in the year 2005, the technological potential and annual energy inflows are shown vertically. The technological potential means possibilities for conversions of the energy inflows into resources with available technologies, and the annual inflow indicates available energy inflow on the Earth that can be captured with the envisaged technologies.

In 2005, about 62 EJ renewable energy is consumed. The minimal technological potential is nearly 140 times larger than all energy consumption that year. About 74% of that renewable energy is consumed from biomass whose technological potential is small compared to the other resources. For example, the minimum potential of solar energy is a few hundred times larger than that of biomass but is hardly used. The annual flows of renewable energy are 17 times larger than the maximum technological potential that is four times the minimum potential. Thus, renewable energy is sufficient for the growing energy consumption of nature and people. Although the consumption cannot grow infinitely, it can increase a few hundred times within a safe margin. The main challenge is the application of renewable energy, not the availability of resources.

Resources of renewable energy are available in all locations on the Earth's surface as solar energy, wind energy, and hydropower. That wide accessibility enables activities on many locations. Their applications are constrained by low energy densities of those resources, measured by the energy content per unit of area or mass. Low energy density implies that a large space is needed per production unit of renewable energy, the so-called 'large footprint'. This causes laborious applications in densely populated areas and energy-intensive activities; for example, in cities. High energy density renewable resources are found only in limited spots and they are too high energy density for their exploitation; for instance, the geyser water and volcanic lava. Furthermore, their density varies over time because depends on changes in seasons, weather conditions, day and night, and others factors that are difficult to manage. This variation, called 'intermittent energy supply', implies that the discontinuities in energy supplies are faced; this is contrary to the continuous deliveries of energy-denser fossil fuels. Therefore, renewable energy is not dispatchable, which means that it cannot be switched on and off unless it is stored. Those properties impede large-scale applications and facilitate spatially dispersed production with storage when energy is abundant, and exchange between producers; for example, during periods of low wind and little sun. For the supplies of intermittent renewable energy, the decentralised production with fine distribution and storage are essential because they enable the balancing of productive supplies with consumptive demands, whilst the present energy systems are developed for the concentrated applications in spatially small spots; for example, in clusters of industries.

The challenging trade-offs between wide accessibility versus large variability and low density require compounding and storage of energy resources. In nature, this is done by the production of plant tissues. Plants have the capacity to capture daylight and convert that light energy with CO_2 from the air into chemical energy through the process of photosynthesis. That chemical energy cumulated during the day is stored as hydrocarbons in the plant tissues which are used for growth at nights when the hydrocarbons are decomposed and CO_2 is released in the process called 'metabolism'. The produced plant tissues, the so-called 'Net Primary Production', is consumed by all organisms (Suneeta et al., 2007). The problem is that the global volume and diversity of biomass declines. It implies that the global availability of this renewable resource declines while production of biomass in particular spots increases. While all man-based biomass increases globally (Peng Li et al., 2017), the global Net Primary Production decreases because larger production of crops does not compensate for losses of virgin forests, and wild animals are decimated by cattle (Bar-On et al., 2018). The shortage of bioresources alongside the biodiversity loss has become a life-threatening issue; this is a threat for many biobased products.

The stored CO_2 in biomass is referred to as sequestration, measured in tonnes of carbon. Per tonne of tree about 0.5-tonne carbon is stored, equivalent of 1.8 tonnes molecular CO_2 in the air; the storage is even larger in soil per tonne tree. When trees are cut and combusted, nearly equivalent of stored CO_2 is released to air along with water vapour and material residues, and a large part of stored carbon in soil is washed out or released to the air. The CO_2 sequestration in 2000 is estimated at 2200 billion tonnes of carbon per year; of which, 20% is in tropical forests, 7% in temperate forests, 26% in boreal forests, and the remaining 47% in various biotopes, of that only 9% in agriculture. The sequestration is on average 62 tonnes per hectare boreal forest (0.8–2.4 tonne tree per ha per annum – t/ha/a); 59 tonnes per hectare temperate forest (0.7–7.5 t/ha/a); and 120 tonnes per hectare tropical forest (3.2–10 t/ha/a). Most of that carbon is stored in the soil, where the carbon density is 1–2 times higher than it is in living plants of tropical forests, and 5–20 times higher in the temperate and boreal ones (FAO, 2001). Thus, the annual carbon storage in the living plants is only 105 billion tonnes of carbon (Beer et al., 2010; Haberl et al., 2007).

As global deforestation is not compensated by afforestation for centuries, net deforestation increases which reduce global renewable energy in biomass, and the sequestration. Below, the scale of sequestration, and net deforestation are estimated for the boreal forests together with the temperate forests, and for the tropical forests, as their properties differ. Trees and soil sequestration are summed up. Estimates of net deforestation in million hectares are compiled based on the state of the world's forests, estimated by the Food and Agricultural Organisation (FAO) of the United Nations (FAO, 2012; FAO, UNEP, 2020), whereas the sequestration is estimated by the multiplication with the average carbon storage. Figure 2.2 shows global net deforestation and CO_2 sequestration; note that the timeline in the Graph is not proportional. The boreal forests together with temperate forests, and tropical forests are in million hectares a year, and the sequestration in billion tonnes carbon per year.

The annual loss of global forests increased from about two million hectares until the mid-1800 to 15 million hectares in the end-1900s, declining by half thereafter. In the 200 years from 1800 to 2000, on average 4.2 million hectares of forest are lost each year, which has reduced sequestration by 0.43 billion tonnes of carbon a year. Until 1900, mostly the boreal and temperate forests were deforested. The loss of tropical forests accelerated until 2000 and declined thereafter. Meanwhile, the afforestation in boreal and temperate forests nearly balanced deforestation. A major driver of that deforestation was the population growth from 1.1 billion people

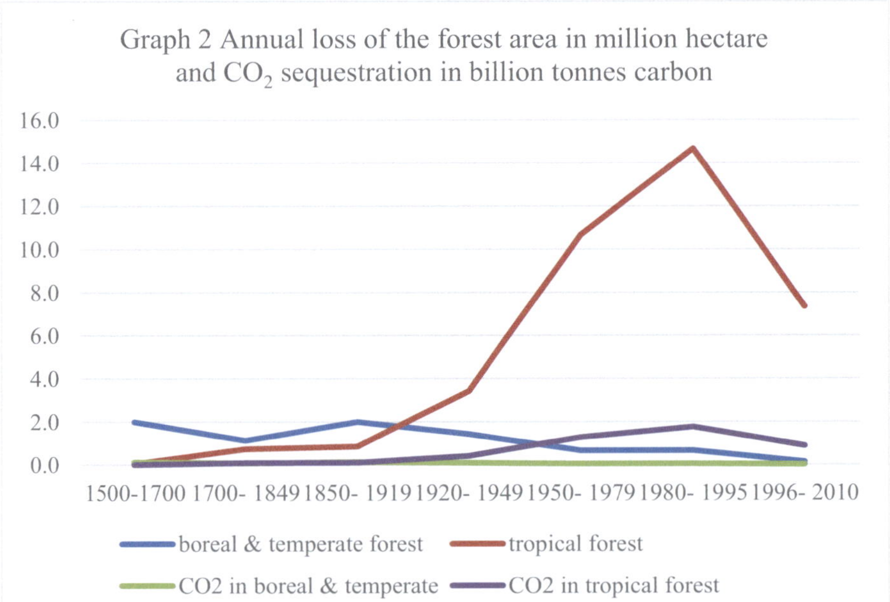

Fig. 2.2 Annual loss of the forest area in million hectare and CO_2 sequestration in billion tonnes carbon

in the mid-1800s to 6.9 billion in 2010 as the deforestation per capita declined throughout the 1900s. The CO_2 sequestration also declined. This accelerated during the second half of the 1900s when more tropical forests vanished. That decline can be turned into an increase of the sequestration if deforestation of tropical forests stops and afforestation of boreal and temperate ones is accelerated. Although technologies for afforestation are available, higher-value services from forests are needed as competition for space is faced. Given that main CO_2 sequestration is in soil, afforestation for the standing forests is needed rather than larger production of timber. Therefore, the challenge is to render higher benefits from the standing forests due to high-value leisure, tourism, education, and suchlike consumptive activities.

It is underpinned that there are sufficient resources of renewable energy to satisfy an energy consumption that is a few hundred times larger than what it is at present, and that technologies for this purpose are available, though low in density and intermittent. It is also shown that a few billion tonnes of CO_2 can be sequestrated in a year if afforestation exceeds deforestation; which is possible if high-value services in forestry are generated and combined with the income from carbon storage.

2.3 Fossil Fuel Resources

Although the flows of renewable energy can satisfy all energy demands, nearly 87% of energy consumption in the year 2005 was based on stocks of fossil fuels. Most stocks are plants fossilised into peat, coal, oil, and natural gas; herewith, data on peat is included in the coal data. The fossilisation emerges when plants die and their tissue degrades. During the degradation of biomass, some carbon is released into the air as methane (CH_4). Most carbon remains as organic carbon in the soil if dead organisms are covered by water and mud. Pressed under thick layers of soil without much air the plant tissues fossilise. After a few million years of fossilisation, the tissues of organisms are converted into peat; a few million years later into lignite; followed by coal, oil, and natural gas in that ascending order of the fossilisation period. This process is driven by longer time, and a larger pressure which corresponds with deeper layers under the soil. Therefore, peat is usually found a few metres under the surface; coal and oil up to a few hundred metres; and natural gas as deep as a few kilometres. Nevertheless, seeps of oil and gas are also found on the Earth's surface. While it takes millions of years before the tissues of dead plants that are not consumed by other organisms are fossilised, technologies aiming at the acceleration of fossilisation failed so far. For example, technologies for the fossilisation of organic waste have been in development for decades, but did not deliver stable products in practice, as yet.

In addition to the hydrocarbon compounds based on biomass, radioactive materials are found on the Earth's surface as a consequence of volcanic eruptions. Among the various radioactive materials, uranium is mainly used as an energy resource in nuclear power plants because it is sufficiently available and radioactive for electricity production after cleaning and enriching of raw minerals from 0.1–0.15% in soil to a few percent for applications. It is also stable for the purpose of energy productions when compared to many other radioactive materials. Large stocks of radioactive materials are also found deep under the sea with volcanic activities. However, technologies for the exploitation of these radioactive materials are in the research phase and the risk of contamination of the sea life by the dispersing of radioactive materials is high.

A difficulty in the exploitation of fossil fuels is that concentrations sufficient for exploitation are available only in particular spots. These spots must be found through costly exploration before the exploitation, while many explored spots are found to be uneconomic for exploitation. If the exploitation starts in such spots, it continues until the stock is exhausted. Improved technologies enable postponement of the exhaustion due to better extraction with chemicals and organisms, as well as drilling into deeper layers of resources on the spot. If a spot is ultimately exhausted, the exploitation is shifted to new sites. Large capital investments are needed for the exploration of valuable spots and their exploitation. Nevertheless, once the exploitation takes off, large streams of capital incomes, which are rents, can be generated during subsequent decades of the exploitation. The generate rents are often manifold higher than the costs of exploitation, which means that high profits are generated.

A mainstream assumption is that the rents increase when more investments for the exploitation of fossil fuels are pursued. Given the flexible allocations of capital for investment, it is argued that a larger demand for capital needed for those investments is reflected in higher interest rates on capital. The growing demand for capital with high profitability implies that the investors demand higher interest rates, entailing costlier exploitation of the stocks. Therefore, it is argued, when fossil resources become scarcer larger investments are needed, the interest rates increase and higher costs impede further exploitation of the scarce stocks. However, the interest rates remained low during last two centuries when the exploitation of fossil fuels increased a lot. For example, the English interest rates – which were decisive during the 1800s and 1900s when England was a global economic power – were 2–5%; except for a short period in 1914 and the late 1900s when they exceeded 10%, but declined thereafter (Interest rates, 2020). Apparently, scarcities of capital for the exploitation of fossil fuels are rarely experienced in practice, with the exception of periods of war when huge unproductive expenditures are generated and a lot of capital is destroyed; not to speak about human suffering. Some scholars argued that the scarcities of capital have not emerged because the resources of fossil fuels are abundant, and many spots are insufficiently explored. Others pinpoint the technological change that compensates for the increasing costs of capital, which covers up the emerging scarcities of fossil fuel resources to cause economic shocks when these resources are exhausted. Whoever is right, the interest rate on capital is a poor indicator of the reserves in the ground. That is why the reserves of fossil fuels are usually based on geological assessments.

Geological assessments indicate possible exploitation of particular fossil fuels, as well as future reserves and technological changes in the exploitation. These complex exercises offer the 'best guesses' about stock in the Earth and are regularly reassessed. The stocks assumed available for exploitation, are called 'reserves' of fossil fuels. Herewith, an authoritative guess based on the Global Energy Assessment is presented. Table 2.2 shows the historical production, the production in 2005, as well as the reserves of fossil fuels that year. The reserves are specified for exploitable and marketable ones, which indicate economic exploitation with current state-of-the-art technologies; and those technically exploitable reserves with future technologies. The reserves are also divided into conventional ones based on the spots explored in the past, and unconventional ones implying a potential for novel technologies, which indicate the exploitable maxima.

About 88 EJ fossil fuels on average per year have been produced globally during the last 200 years, which increased to 416 EJ in 2005. This one is composed of 23% coal, 38% oil, 23% gas, and 6% uranium. Given the growth in energy consumption, the maximum reserves of conventional fossil fuels, excluding coal, are estimated to suffice for several decades; which can be extended for a few centuries when the unconventional resources are included. Nuclear energy is often promoted as the alternative, backstop technology. However, the reserves of radioactive materials for nuclear energy cannot support even present total energy consumption for longer than several decades because the uranium reserves are too small, even when unconventional reserves are considered. Other radioactive materials are also explored and can be exploited, but this energy production is still in the research phase; for example, the occurrence of natural gas and radioactive resources is large, though difficult to exploit because located in deep seas. Coal is the only exploitable fossil fuel that is sufficient for several centuries because its reserves exceed all other reserves of fossil fuels put together. Coal enables larger energy consumption over a few thousand years, but its production and combustion generates severe environmental impacts, in particular high CO_2 emission per tonne, thereby large impact on the climate change per energy unit. Therefore, policies in several countries aim to phase out coal, despite large reserves.

Stocks of fossil fuels are available. However, they are expected to be nearly exhausted within several generations taking the growing level of energy consumption into consideration; except large reserves of coal. This perspective urges a global shift from fossil fuels to renewable energy.

2.4 Shifts in Energy Resources

Nearly all those resources were known for centuries but hardly used. For example, biomass was processed into charcoal; biogas was used for heating and cooking; hydropower replaced work of animals and humans; wind energy generated motion; geothermal energy served in cooking and spas; solar energy was amplified for

Table 2.2 Assessment of fossil fuel reserves based on the Global Energy Assessment

	Historical production		Exploitable and marketable reserve		Exploitable technical reserve			
In EJ = 10^{18} J; 1 EJ ≈ 278 TWh = 10^{12} Wh	till 2005	2005	Minima	Maxima	Minima	Maxima	Occurrent on earth	Years available maxima
Oil conventional	6069	148	4900	7610	4170	6150		42
Oil unconventional	513	20	3750	5600	11,280	14,800	40,000	
Oil subtotal	6582	168	8650	13,210	15,450	20,950	40,000	125
Gas conventional	3087	90	5000	7100	7200	8900		99
Gas unconventional	113	10	20,100	67,100	40,200	121,900	1,000,000	
Gas subtotal	3200	100	25,100	74,200	47,400	130,800	1,000,000	1314
Coal	6712	124	17,300	21,000	291,000	435,000		3514
Uranium conventional	1218	25	2400	2400	7400	7400		300
Uranium unconventional	34	0	0	0	7100	7100	2,600,000	
Uranium subtotal	1252	25	2400	2400	14,500	14,500	2,600,000	587
Total conventional	17,086	386	29,600	38,110	309,770	457,450	0	1184
Total all	17,746	416	53,450	110,810	368,350	601,250	3,640,000	1445

warming; electricity in animals was used for medical purposes; peat, coal, oil, and natural gas were exploited for heat and steam; nuclear materials were known as a hazard, and so on. However, the annual energy consumption was only 10–12 kWh per person in agricultural societies, spanning a few thousand years. This doubled to 20–25 kWh per person in Europe and about 20% higher in Asia during the early industrialisation of the early-1700s. The latter is similar to the energy poverty in the present day. During the last three hundred years, global energy consumption per person increased about 200 times to average of 10,000 kWh per person in the year 2000. It grew in Western Europe, North America and Oceania; followed by Japan, and Russia; and thereafter in Asia, Africa, and Latin America. Corresponding with the growth in consumption, differences across countries increased. For example, the average consumption per capita in the USA was five times larger than the global average in the year 2000, whilst the Latin American average was 70% of the global average; however, the Asian and African ones were only about 30% of Latin American while two-thirds of the global population lived on these two continents. This growth evolved alongside changes in the resource composition of energy consumption throughout two centuries; the drivers of energy consumption are discussed in the next chapter. These changes offer insight, which helps in assessing assess possibilities for the shifts from fossil fuels to renewable energy in the future.

Larger energy consumption evolved along with the global shift from renewable energy to fossil fuels. This is shown in Fig. 2.3 based on OWD data, which covers the energy resources in TWh from the year 1800 to 2000; herewith, 'geoth' means geothermal energy. This data excludes informal energy resources, and the labour of animals and people based on muscles whilst their share in all energy resources was about 1% in the USA, 1.8% in Canada, 2.7% in Germany, 4.2% in Italy, and higher

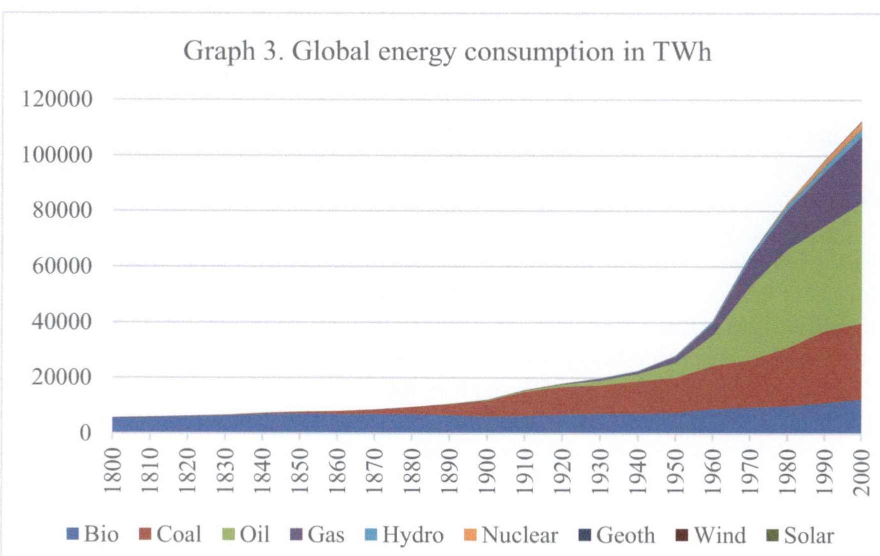

Fig. 2.3 Global energy consumption in TWh

in other countries in the 2000s (OWD). These were higher shares than the share of modern renewable energy, but their growth was close to nil.

Global energy consumption multiplied 20 times over two centuries – from 5.6 PWh (20.3 PJ) in 1800 to 112 PWh (405 EJ) in 2000 – mainly due to the use of fossil fuels from the mid-1800s onwards. Whilst bioenergy rarely grew faster than 0.4% a year throughout that period of time, coal grew 2.4% from the mid-1800s to 1900; the consumption growth of coal slowed thereafter. Coal is followed by twice higher growth rates of oil and gas from 1900 to 1950. As result, fossil fuels covered only 2% of all energy resources in the early 1800s, which increased to 72% in 1950 and 86% in 2000 when the global growth of energy consumption and fossil fuels had saturated. Hydropower grew after 1910 and nuclear power after 1950, but the consumption of each one reached only 2% of the total energy consumption in the year 2000. Geothermal energy emerged in the 1920s and grew slowly, whereas wind and solar energy grew fast after the 1980s, though their total share until 2000 remained below 1% of the global energy consumption, and the biofuels emerged on a small scale in the 1990s. Those four resources are considered modern renewable energy.

The emergence of fossil fuels in Western Europe has been explained by high prices of bioenergy from wood, because of depleted forests along with growing demands for energy in industries, which is met by the introduction of coal (Cipolla, 1993). This mainstream argumentation about wood scarcity as the driver of innovation can explain the emergence of fossil fuels in the late 1700s in Western Europe. The emergence and growth of fossil fuels during the 1800s in the USA was not solely driven by scarce bioresources. An explanation in line with evolutionary argumentation is that expanding industries demanded energy denser resources than biomass, and policies enhanced fossil fuels through taxes, infrastructure, and institutions; for example, the taxation of biofuels. This is also not the full story because fuelwood was used in many industries for combustion, and for gasifiers in vehicles well into the 1900s. An additional explanation with reference to behavioural thinking about innovations is that fossil fuels were also easier to handle than wood and coal in production, transport, and consumption. Presumably, a cumulation of those factors invoked a fast shift to fossil fuels in the late 1800s which is discussed in more detail in the next chapter.

The global shift from traditional renewable energy to fossil fuels evolved in fifty years when the USA and European economies grew fast. While the growth of renewable energy saturated, all fossil fuels grew tenfold in the late 1800s when mineral oil grew 150 times from 1870 to 1920. The consumption of fossil fuels accelerated again during the economic boom in the mid-1900s after the 2nd World War and reached more than an 85% share by the 2000s. However, this data about the global shift to fossil fuels can be biased by data about the USA and Europe. There is less historical data on Asia and Africa where traditional renewable energy is still widely in use. Furthermore, coal remains a key fossil fuel on all continents except the Americas that use mainly oil. From the late 1900s on, wind and solar energy expanded at two-digit growth rates. This expansion suggests a shift from fossil fuels to modern renewable energy whilst it can also replace traditional renewable energy.

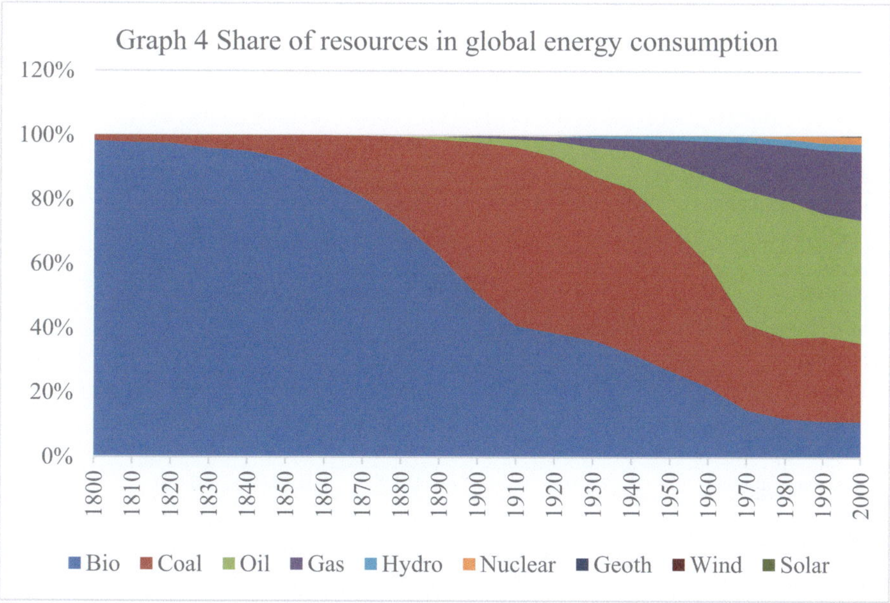

Fig. 2.4 Share of resources in global energy consumption

Although fossil fuels rivalled traditional renewable energy, and modern renewable energy emerged as the rival of fossil fuels, substitutions on the global scale were rarely observed. Nearly all energy resources grew throughout the last centuries. The vested energy resources and technologies were complemented by novel ones whose growth rates remained higher over decades. This complementation mechanism evolved during the growing energy consumption. As a result, the resource composition of energy consumption changed and diversified during those two centuries, as shown in Fig. 2.4 using the OWD data.

Biomass was complemented by coal in the late 1800s; then oil and hydropower complemented coal from the early-1900s onwards; followed by gas and nuclear power in the mid-1900s; and finally, geothermal, wind energy, and solar energy after the 1980s complemented all previous resources. This complementation mechanism continued throughout two centuries when nearly all resources grew in total, even ones based on animal and human labour, as well as the oldest fossil fuel that is based on peat; however, their consumption per person declined.

This complementation in energy resources from 1800 to 2000 is illustrated with the USA data (O'Connor & Cleveland, 2014). During these two centuries, the USA grew into the largest economy and energy consumer. Table 2.3 shows economic indicators and changes in energy resources from the years 1800 to 2000, in fifty-year intervals. It includes non-traded biomass, and calories that are consumed in foods for animals and human labour. Throughout the last two centuries, the population of the USA grew faster than the global population. Its population growth evolved alongside a faster growth of income per capita than energy consumption per capita.

Table 2.3 Energy resources in the USA from 1800 to 2000 based on O'Connor and Cleveland (2014)

	1800	1850	1900	1950	2000
General indicators					
Population millions	6	24	76	152	283
GDP in USD_{2011} billion	11	67	478	2321	12,874
Energy consumption in TWh	197	904	4450	11,814	28,238
Energy per capita, MWh/Person	25	36	58	77	100
Energy-intensity, kWh/USD_{2011}	13	13	9	6	2
Share of energy resources in consumption					
Muscles	20%	10%	7%	2%	1%
Biomass	78%	75%	16%	4%	2%
Wind	1%	0.5%	0.1%	0.001%	0.1%
Water	0.2%	0.2%	0.1%	1%	1%
Other renewables	1%	0.1%	0.1%	0.0%	0.0%
Coal	0%	7%	45%	31%	23%
Petroleum	0%	3%	19%	35%	39%
Natural gas	0%	3%	12%	12%	24%
Nuclear	0%	0%	0%	15%	9%

The *energy-intensity* of this economy declined fast from the late 1800s on. All renewable energy resources increased, except local uses of geothermal energy, natural ice, and a few other small-scale resources that stagnated. Despite small scale, those local energy resources did not vanish; even natural ice is sometimes transported for consumption in posh restaurants. The growth of fossil fuels accelerated in the late 1800s. Fossil fuels expanded in the USA from 1850 to 1900, when they increased from 14% to 66% of all energy resources whilst energy consumption grew 15 times. That shift in the USA was a major driver for the global shift to fossil fuels in high-income countries during the late 1800s, followed by mid-income and low-income countries in the late-1900s when modern renewable emerged and grew fast, while fossil fuels grew as well.

The fast growth of novelties can be comprehended by lower prices and superior qualities. The reasons for maintaining the inferior resources and technologies from the past do need an explanation. The complementation of energy resources implies that the consumption of incumbent products remains for many decades, possibly even centuries. This consumption can be interpreted as hoarding, it means in economics keeping large stocks as a reserve for the periods of insecure supplies; for example, during economic crises or wars. This is often associated with speculation, the intention to profit from the expected scarcities in the future. However, hoarding is an insufficient explanation of the persisting complementation in the energy resources. Fear of scarce energy resources did trigger storage facilities for fuels, but these facilities were small scale compared to production. While the international trade in fossil fuels grew, these facilities did not exceed 1% of the global production during last decades. A more convincing explanation addresses the backward and forward linkages of technologies.

The backward linkages refer to the dependency on suppliers in value chains. Generally, the suppliers impede changes in value chains because confront high costs of changing equipment, management procedures, and other costs of change-over, as well as the risks of incompliance with the novel interests. They reinforce the vested interests when their appeals to authorities for the maintaining of status quo are approved in regulations that impose the so-called 'barriers of entry'. These barriers of entry refer to tariffs and non-tariff regulations; for example, excise duties on imported fuels and certificates for local biofuels. The forward linkages refer to specialisations of the incumbent interests in services for market niches; meaning value addition by services tuned to particular consumers. For example, coal-based steam engines were replaced by oil-based internal ignition in moving machines (movers) and vehicles, yet coal consumption kept growing in stationary power plants. If the specialisations continue, they diversify technologies whilst every specialist improves its performance, thereby it is more difficult to introduce competitive novelties; e.g. coal consumption becomes more energy efficient despite lower energy density than oil and gas. The complementation is observed when the global population increased from 0.8 billion people in 1800 to 4.7 billion people in 2000, and the average GDP per person in USD_{2011} grew from about 1000 to more than 17,500 a year. Contrary to the data on economic growth, the data on long periods of economic contractions are unavailable. It implies that it is not possible to verify the substitution or complementation in energy consumption during the economic decline.

Presently, modern renewable energy is a rival of fossil fuels and traditional renewable energy from the past; however, the incumbents are rarely replaced. The rivalry between energy resources alongside complementation continues as shown in Table 2.4 with global consumption of fossil fuels, traditional renewable energy, and modern one in 2015 and changes during 1990–2015 based on IEA data.

Fossil fuels and renewable energy can be considered rivals while both grew from 1990 to 2015 at various growth rates globally and across countries. While fossil fuels grew fast in all mid-income and low-income countries they hardly grew in the USA, declined in Japan, the EU, and Russia. All renewable energy was about 14% of all energy consumption whilst the traditional ones based on biomass and hydropower covered nearly 90% of that. It is the main energy resource in many mid-income and low-income countries; for example, 94% in Ethiopia where hardly any fossil fuels are produced, as well as 80% in Nigeria where large-scale production of fossil fuels is found. Modern renewable energy covered about 1.5% of global energy consumption in 2015; the consumption of geothermal energy was higher in the Philippines (18.4% of all energy), Indonesia (7.7%), and Mexico (2.2%) which are volcanic active areas. Generally, modern renewable energy was consumed mainly in high-income countries (2.9% share in the EU, 1.5% in Japan, and 1.4% in the USA); and a few mid-income ones (1.6% in China, 0.9% in Brazil, and 0.6% in India); it hardly grew in other countries. Traditional renewable energy grew slowly in nearly all countries whilst modern one grew at a two-digit rate in mid-income countries and the EU, but hardly in the low-income countries. Therefore, the resource composition of energy consumption changes fast across countries.

Table 2.4 Consumption of fossil fuels, traditional renewable energy and modern one in TWh and annual average growth rate from 1990 to 2015 based in IEA data

	Fossil fuels		Traditional renewable energy		Modern renewable energy	
In TWh, annual average growth	2015	1990–2015 (%)	2015	1990–2015 (%)	2015	1990–2015 (%)
World	137,092	1.8	19,281	1.7	2333	7.1
United States	24,280	0.6	1399	1.4	364	3.4
Japan	4701	−0.2	217	2.2	73	3.1
European Union	15,814	−0.6	2079	3.7	530	11.5
Russian Federation	8008	−0.8	256	−0.7	2	11.2
Mexico	2010	2.0	130	0.5	49	0.3
Brazil	2034	3.6	1364	2.3	30	20.9
China	31,610	5.6	2434	0.0	537	37.4
Indonesia	1746	4.3	675	1.2	201	9.1
Philippines	385	4.4	109	0.4	112	3.0
India	7421	5.5	2421	1.6	56	29.1
Nigeria	317	2.7	1304	3.1	0	0.0
Pakistan	677	4.0	415	2.3	1	0.0
Bangladesh	330	6.6	108	1.2	0	0.0
Ethiopia	35	5.9	542	3.0	0	0.0

The growth of fossil fuels and traditional renewable energy saturated, whilst the growth of modern renewable energy accelerated in high-income and a few mid-income countries. While the rivalry between fossil fuels and renewable energy is widely acknowledged the rivalry between the traditional renewables and modern ones is also observed particularly in mid-and low-income countries. The latter has important consequences for energy access and climate change. To illustrate this importance, a simple simulation of the past trend is made. Had the traditional renewable energy grown as fast as modern renewable energy while substituting fossil fuels during 1990–2015, all renewable energy would cover about 50% of all energy resources, CO_2 emissions would decline and climate change would be mitigated. Therefore, the growth of all renewable energy resources contributes to the mitigation of climate change whereas too selective policies impede that. For example, the fussy debates about bioenergy versus food impede the use of bio-residues for renewable-based heating and biofuels.

Renewable energy grew and diversified during the late 1900s. Although modern renewable energy is still small-scale compared to fossil fuels and traditional renewable energy, it grows fast albeit the rates substantially differ across the modern resources. Figure 2.5 shows the production of modern renewable energy in TWh, from 1950 to 2010, in ten-year intervals with SHIFT data.

Geothermal energy emerged in the 1920s and grew steadily. Biomass and biowaste were converted into sugars and cellulose for distillation into bio-methanol and bioethanol from the 1970s onwards in the USA and Brazil; vegetable oils were refined into biodiesel in Europe after the 1990s, and biomass

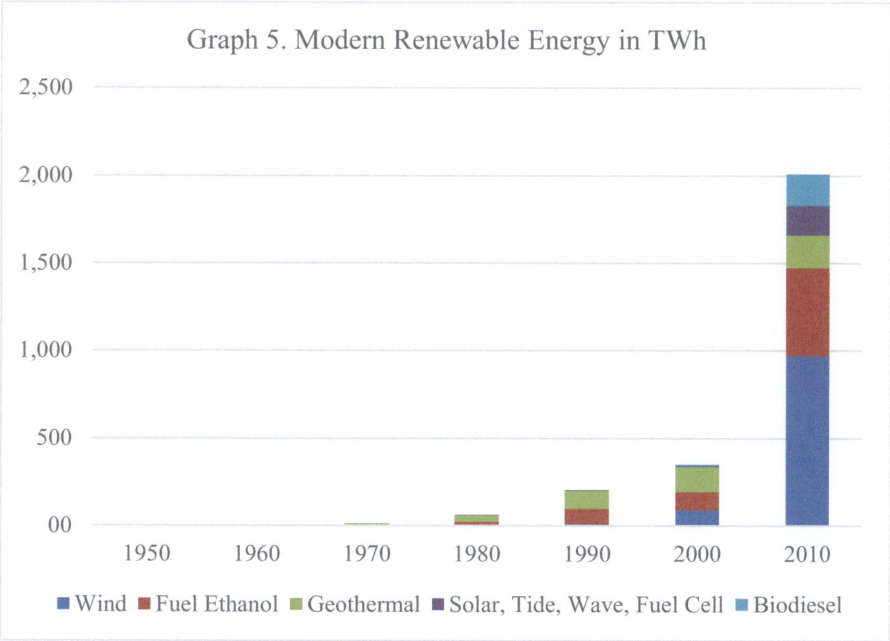

Fig. 2.5 Modern renewable energy in TWh

residues were digested for biogas in Europe and Asia; all those are related to farming. Wind energy grew globally from the late 1980s on, and solar energy from the late 1990s; followed by local introduction of tidal and wave energy. High fuel prices triggered this introduction while the responses to the prices differed across the continents, even more across countries. When the fuel prices increased from 1974 to 1985, the EU production was surpassed by the USA one, followed by the Latin American production, in particular in Brazil. During low oil prices in the 1990s, modern renewable energy experienced a setback in the Americas, whilst the production in Asia expanded and the European one steadily grew. When oil prices increased again from 2005 to 2015, the production grew fast in Asia and Europe, less in North America and hardly in Latin America, and Oceania. Meanwhile the production of renewable energy in Africa hardly took off. After the decline of oil prices in 2015, the production kept growing in Asia and Europe but slowed down on other continents.

Growing electricity production was an important driver for that growth. Figure 2.6 shows the indices of global income in USD_{2011} based on Madisson. Energy consumption is based on OWD. The electricity production, as well as all renewable energy and modern one, are based on the UN from 1930 to 1950 combined with the BP and World Bank (WB) data from 1960 to 2010. All energy is in indices of TWh. All those data are shown in intervals of ten years; the electricity in 1960 is interpolated because many data between the years 1950 and 1965 are missing.

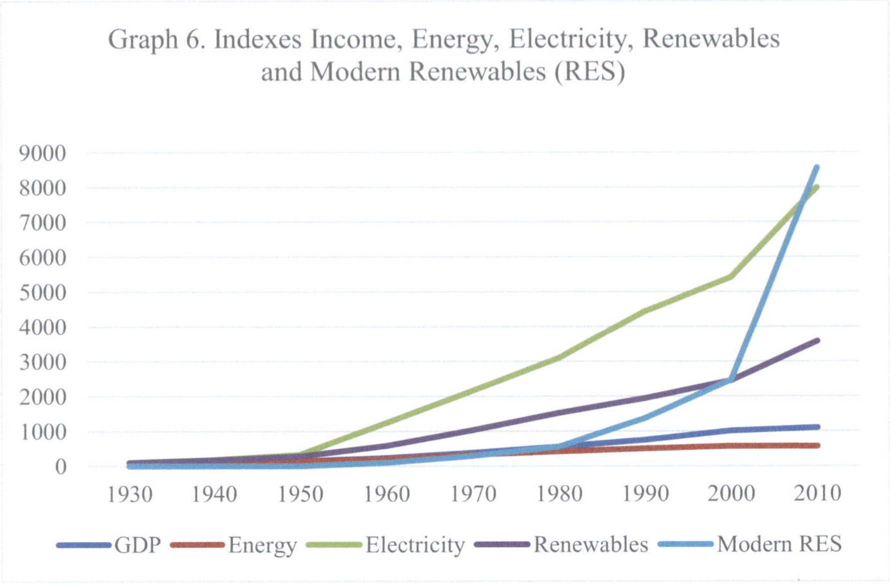

Fig. 2.6 Indexes income, energy, electricity, renewables and modern renewables (RES)

While electricity production grew faster than income and all energy throughout the 1900s, modern renewable energy as a resource for electricity generation grew even faster; these were mainly wind and solar energy. The electricity generation emerged in the USA in the early 1900s, soon followed by Europe and a few decades later by Russia at even faster growth rates. It was initially based on hydropower but thermal production gained a share. Renewable energy covered about 44% of electricity production in 1930, by and large hydropower whose share declined to 20% by 2010. The global production of electricity expanded during the economic boom after the 2nd World War- even accelerated in the 1990s – while hydropower did not keep up the pace, though considered the cheapest power resource. Thermal power grew faster in the 1950s followed by nuclear power that stagnated after the 1980s because it was considered costly and risky. Compared to those resources, modern renewable energy emerged as late as the 1980s. It grew faster than electricity production but its share in all electricity resources did not exceed 8% until 2010, whilst the use for heating is limited to biofuels. Electrification generated innovative services in fine mechanics, data processing, cultural expressions, and many other applications. By the late-1900s, the services based on electrification of economies generated a large share in GDP and became a vehicle for the growth of modern renewable energy by the late 1900s (Griffith & Calish, 2020).

Decisions about fossil fuels versus renewable energy refer not solely to the technical scarcities in nature but also to the normative choices in economic development. Usually, the decision-making drives conversions towards energy denser resources and products; for example, denser syngas from coal, hydrogen from gas, and nuclear energy from uranium. These are useful for the energy-intensive industries but hardly valuable in service and on the contrary, they pose safety risks and cause pollution. Meanwhile, the wide availability of renewable energy enables value addition in energy services in the consumers' networks. As the energy-intensive industries constitute a diminishing share of GDP whilst growing knowledge capabilities contribute to higher-value services, the distributed energy systems can generate welfare at somewhat higher costs of energy resources. The societal choices between higher efficiency due to energy-dense fuels versus broader access to valuable services with renewable energy evolve in favour of the latter option.

The complementation of energy resources causes barrier for the energy transition, Modern renewable energy received an impetus from high prices of fossil fuel in the 1980s and early 2000s, while the growing electricity generation was an important vehicle for its growth and diversification. The value-addition by energy services evolves toward the driving force in energy consumption.

2.5 Growing Pollution

Combustion of fuels causes pollution issues to exacerbate from a local to global scale. Explosions and fires in coal mines, as well as in the production and distribution of oil and natural gas cause accidents at the workplace and their spills threaten nature. Locally, imperfect combustions of dung, biomass, peat, and coal emit toxic carbon monoxide with casualties in badly ventilated houses. Although conversions of biomass into cleaner charcoal are known investments in energy-efficient stoves and purchases of charcoal are costly for low incomes. Effectively, many people keep obsolete stoves in households despite severe respiratory problems, which is a major issue in households of low-income countries. The combustion of coal, oil, and gas in the power plants and at households improve energy efficiency compared to cheaper dung, wood, and peat; however, as energy consumption grows, more pollution is generated and dispersed over wider distances. In effect, the combustion of coal emits particles that cause smog, a major cause of lung diseases. Further, water for cooling the power plants undermines local ecosystems because released as hot, dirty water. Combustion of fossil fuels also emit SO_2 and NOx which are toxic, cause acidification of nature, and degrade materials on the regional scale, whilst burning gasoline and diesel in transport emits NO_x and fine particles which are toxic compounds causing cardiovascular and respiratory diseases and cancer.

Those local and regional issues are being partly resolved in high-income countries since the late 1900s, due to better conversions and pollution control technologies that are enforced by regulations. However, nature and people in mid-income and low-income countries still suffer from those pollutants. While technologies are available, they are not enforced because regulations are lax. Particularly, living in

African, Asian, and Latin American metropolises is unbearable for people whose health is sensitive to air pollution. As a consequence, about eight million people die yearly because of airborne pollution, which is 18% of all global deaths (Burrows, 2021). This death toll does not include diseases and deaths caused by deficient water, sanitation, and polluted soil related to heat and power production. It is also estimated that fine dust ($PM_{2.5}$) from the combustion of fossil fuels causes yearly about 10.2 million premature deaths, mainly in China and India (Vohra et al., 2021). However, it is also pinpointed that the health impacts of pollution caused by the combustion of fossil fuels is only a fraction of the impacts caused by indoor pollution when traditional biomass resources are burned and dangerous habits as smoking cigarettes persist (Markandya & Wilkinson, 2007).

While many people suffer from those local and regional issues that remain unresolved despite low-cost cleaner technologies, pollution by greenhouse gases dominates political agendas with regard to the risks posed by climate change. The main global issue is the large volume of CO_2 emissions. Global CO_2 emissions multiplied nearly 750 times, from 31 million tonnes in the year 1800 to more than 23 billion tonnes in 2000. This grew alongside energy consumption that multiplied 20 times, mainly because of greater fossil fuel combustion. Fossil fuels cause CO_2 emissions along with residues of water vapour and carbon-rich ash as the hydrocarbons in the fossilised plants to deliver heat during combustion. Heat is the targeted quality. Vapour is a by-product that can be used for steam in engines. It is often emitted while considered cheap and harmless; however, steam releases cause injuries and vapour acts presumably as a greenhouse gas at a height of a few kilometres where aeroplanes emit vapour caused by combustion of kerosene. Ash is usually disposed of, which causes toxic leaks into soil and water, but it can be immobilised for re-use in building materials. Nearly all CO_2 is emitted considered useless.

The emission of CO_2 is called a greenhouse gas because it impedes the outflows of reflected solar energy from the Earth into space when it accumulates in the atmosphere instead of sequestration in oceans, soils and trees. Thus, deforestation aggravates pollution because it diminishes the sequestration of CO_2 in living plants, and soil. The accumulation of greenhouse gases causes climate change. This is manifested as higher and fluctuating temperatures of the air and seas, heavier evaporation and precipitation, more frequent wind storms, acidification of oceans and other hazards. From the late 1980s onwards, the concentration of CO_2 in the atmosphere increased fast. It fluctuated steadily between 200 and 300 parts per million (ppm), throughout many centuries and increased to 400 ppm from the 1990s to 2000s; going up to even higher thereafter. The increasing combustion of fossil fuels during the last centuries became the largest source of CO_2 emissions; CO_2 emissions are also generated in the production of cement and fertilisers while their scale is small compared to the combustion of fossil fuels. By the 2000s, those annual man-made CO_2 emissions were larger than all non-man-made emissions put together, from eruptions of volcanos (Scientific American, 2020), to the emissions from fires, plant metabolism, and other natural causes (Brahic, 2007). Presently, man-made CO_2 emission is the main cause of climate change and by far the largest greenhouse gas by mass; thereby it causes the largest total greenhouse impact.

Methane (CH_4), fluorocarbons (HFCs and CFCs), laughing gas (N_2O) and a few other small-scale emissions pose stronger hindrances for the reflected solar flows from Earth to space, persisting longer in the atmosphere per unit of mass, but their total mass is small. Therefore, all greenhouse gases are expressed in the CO_2 equivalent, referred to as 'CO_2 emissions' for convenience.

Although the greenhouse effect of CO_2 on climate change has been known for more than a century, first discovered by Swedish chemist Svante Arrhenius in 1896, it was issued on international political agendas as late as the 1980s when the International Panel for Climate Change (IPCC) was established to provide scientific support of political decisions. The incubation period of the policymaking – from signalling of this environmental issue in science to the introduction of the global policy – spanned nearly 90 years. This global policy was set by the first international agreement on CO_2 emissions reduction within the United Nations Framework Convention on Climate Change in 1997, called the 'Kyoto Protocol' because it was arrived at in Kyoto (Japan). After 1997, international policy instruments were introduced and ratified by most countries; including instruments for emission trading, financial compensations for emission reduction, and financial support of pollution reduction technologies for low-income countries. Two decades later, the international agreement on climate change in Paris was reached, which aims at 350 parts per million (ppm) CO_2 in the atmosphere. However, this agreement lacks policy instruments because it is based on the voluntary monitoring of actions of individual countries; however, even uniform measuring method, monitoring and transparent reporting are lacking.

Whenever coal, oil, and natural gas are combusted in life cycles of the energy service CO_2 is emitted because combustion involves the reaction of hydrocarbons in fossil fuels with oxygen in the air. Gasification and pyrolysis of fossil fuels are conversion technologies that produce syngas with little use of oxygen, which largely avoids CO_2 emissions, but they can emit methane. Table 2.5 shows the CO_2 emission factors in gram per kWh fossil fuel and electric power based on a few authoritative data sources. The first column shows the energy resources. The second and third columns show the CO_2 emission factors in 2015 based on combining of IEA data on global CO_2 emissions per fossil fuel with BP data on fossil fuels (IEA-BP), and an annual average change in emission factors when those emissions are annually divided by energy resource during 1990–2015 (IEA-BP). In the fourth and fifth columns, these are combinations of IEA global CO_2 emissions and IEA energy resources (IEA-IEA). The next two columns show CO_2 emission factors of electricity production based on IEA combustion data (IEA, 2019), and IPCC compilation of many life cycle assessments based on energy resources (IPCC, 2012).

Differences in the emission factors can be observed. The IEA-BP data show lower emission factors than the IEA-IEA data, and the changes differ; for example, the emission factors of natural gas decrease in the IEA-BP data, but they increase in the IEA-IEA data. The emission factors of electricity are lower in the IEA compared to IPCC. The World Bank data and its growth rates differ somewhat from the

Table 2.5 Emission factors of energy resources

CO_2 gram/kWh annual average change '90–2015	CO_2 – IEA, energy BP		CO_2 – IEA, energy IEA		Electricity production	
	2015	Change	2015	Change	IEA	IPCC (1)
Coal (2)	320	0.8%	1140	0.8%	845–1020	990 ÷ 350
Oil (3)	230	−0.02%	250	−0.5%	600–675	800 ÷ 250
Natural gas (4)	190	−0.8%	350	0.6%	400	480 ÷ 100
Nuclear	–	–	–	–	–	10 ÷ 50
Biomass	–	–	–	–	–	30 ÷ 50
Hydropower	–	–	–	–	–	Nil
Geothermal	–	–	–	–	–	50 ÷ 50
CSP (5)	–	–	–	–	–	20 ÷ 10
Photovoltaics	–	–	–	–	–	50 ÷ 50
Wind energy	–	–	–	–	–	20 ÷ 20
Global mix	0.27	0.03	0.65	0.2		
World Bank mix	0.2	−0.45				

(1) median and 75th percentile of life cycle assessments; (2) Peat is below the lowest value of coal and lignite is similar to the highest value of coal (Jurich, 2016) (3), Shale oil emission factor is 1195 gram per kWh, (4) Blast furnace gas is 2430 gram per kWh (5) Concentrated Solar Power

IEA-BP and more from the IEA-IEA data. The causes of those differences are not assessed because they are not essential for the purview of this book, but they are important for the monitoring of international agreements. The IEA emission factors are similar to the combination of IEA-BP data shown in column 2, which are used below for the estimate of changes in the global CO_2 emissions.

In practices, the CO2 emissions are rarely measured at the emission sources. The estimated CO_2 emission factors per resource are mainly based on the chemical compositions of fuels along with assessments of the combusting conditions whereas measurements of CO_2 in the vent during combustion are rarely done. The hydrocarbons composition of all fossil fuels can be formalised by the chemical formulas: $C_nH_{(n+m)}$ for $m > n + 2$ for coal; C_8H_{18} for oil; and CH_4 for natural gas. Based on these formulas, the emission factors show a 28% higher emission of coal than oil, and 18% higher oil than natural gas. The electricity consumption of all resources shows CO_2 emissions. Renewable energy has 10–40 lower emissions than gas, hydropower has nil emissions according to the IPCC though masses of concrete and other energy-demanding materials are used for construction and maintenance which are energy-intensive activities. IPPC also shows low emission factors for the nuclear power but estimates that include upgrading and end-of-life indicate a few times higher emission factors. Large differences in the emission factors across life cycle assessments of energy resources for electricity can be noted. Despite those differences, it is beyond a doubt that there are substantially lower CO_2 emissions of renewable energy and nuclear power when compared to fossil fuels. Note that the greenhouse impact of coal gasification in electricity generation is also several times larger compared to renewable energy (Ruether et al., 2004).

Those differences in the emission factors of fossil fuels do not justify preferences for a particular fuel because lower carbon content in the chemical composition of fuels can be countered by the inferior qualities of resources, lower energy-efficiencies in combustions, deficient conversions in the life cycles of suppliers and spills of resources. This is relevant with regard to arguments purporting that natural gas is 'clean' and coal is 'dirty' with respect to the phenomenon of climate change. It can also be noted that the emission factors of coal increased during the last 25 years, whilst the emission factors of oil and gas decreased; which can be caused by lower coal quality, deficient data, and presentation of the most favourable data by organisations that deliver and check this data. Assessments of biases in the use of data by authorities are not found; nevertheless, biases can be expected. Regarding differences in the emission factors shown in Table 2.5, scrutiny of data and uniformity in measurements are needed for reliable policymaking.

The CO_2 emissions caused by fossil fuels are estimated through the multiplication of the IEA-BP emission factors in 2015 with BP data on energy resources; changes in the emission factors are ignored. Figure 2.7 shows global CO_2 emissions from 1965 to 2017.

The CO_2 emissions grew steadily during the last 52 years. This estimate shows 33 billion tonnes global CO_2 emissions in 2015, compared to the IEA estimate of 32 billion tonnes and the World Bank's estimate of 36 billion tonnes, which make for small differences; a billion tonnes CO_2 is often written as gigaton or $GtCO_2$. The

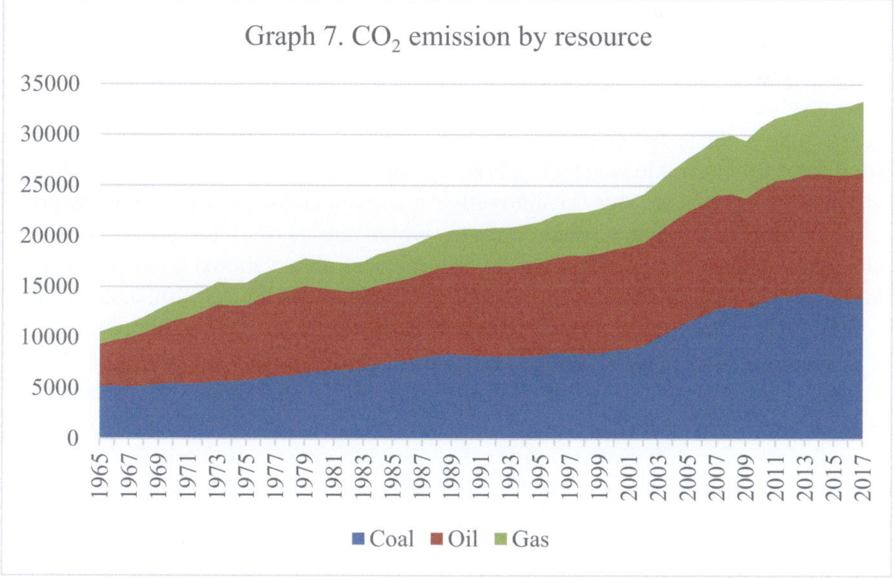

Fig. 2.7 CO_2 emission by resource

growth of CO_2 between 1965 and 2015 was a 2.3% annual average, a statistic of continuous growth that is often pinpointed in the media. Nevertheless, for the projections of CO_2 in the future, it is relevant to assess whether the growth trend is steady, accelerated, or saturated. Herewith, it is assessed that the growth has saturated from 2000 to 2015. This saturation is caused by several factors. One factor is the changing composition of fossil fuels, because 43% of all CO_2 emissions in 2015 were caused by coal, 37% by oil, and 20% by gas when the growth rate of coal was only 1.9%, compared to 2.2% of oil and 3.5% of gas throughout those forty years. The growth of coal and oil has saturated, whilst the growth of gas has accelerated. Another factor is that the growth of global energy consumption is saturated. The third factor is the acceleration in the growth of modern renewable energy.

Until 1990, less than 21 billion tonnes of CO_2 was emitted per year, which is reflected in the concentration of CO_2 in the atmosphere that fluctuated within the band of 250–300 ppm. This concentration is considered acceptable for the greenhouse effect because CO_2 in the atmosphere is also absorbed on Earth; for example, the ocean is an important sink. However, sinks are gradually saturated. Compared to 1990, an additional 12 billion tonnes of CO_2 was emitted in 2015, which adds to the saturation of the global sinks. If the growth rates during those 25 years from 1990 to 2015 are extrapolated, about 73 billion tonnes of CO_2 emissions can be expected by the year 2050. By then, a minimum of 73% emission reduction is needed in order to attain the emissions of the year 1990 with no consideration about the saturation of global sinks in the future. Therefore, the reduction in emissions of global CO_2 by more than 73% in about 30 years is needed to maintain the availability of the environmental qualities that mitigate further climate change. This CO_2 emission reduction on the global scale within one generation should be considered mandatory, which should be pursued as fast as possible for the prevention of severe impacts of further climate change. Given the growing demands for energy, this target for emission reduction cannot be attained solely with higher energy efficiency. Growth of renewable energy that replaces fossil fuels is needed.

An issue is the speed of changes in consumption. The duration of the shift from fossil fuels to renewable energy is a bone of contention. Some experts expect a shift within several decades due to innovative energy technologies and a widespread sense of urgency about climate change, which are needed for the adoption of technologies (Sovacool, 2016). Based on historical observations, however, there is doubt about this possibility with respect to sluggish innovation diffusion, because technologies from the past are persistent (Grübler et al., 2016), incumbent interests obstruct changes (Fouquet, 2016), and meagre results in the mitigation of climate change despite ambitious plans (Smil, 2016). Both viewpoints could hold true because past shifts in the consumption of energy resources evolved over centuries. Nevertheless, the shift from traditional renewable energy to fossil fuels evolved within the period of 50 years of the late 1800s and early 1900s, when fossil fuels captured about two-third of energy consumption. It indicates that changes can accelerate over short periods of time. Whoever is right, more worrying is the complementation process in growing economies. More troubling is the observation that the energy resources are usually complemented by new ones while rarely replaced because of the backward and forward linkages between interests and technologies in the value chains that are vested in the past. How to overcome the impediments posed by these incumbents is a major political challenge.

Perseverance in policies is needed for more than 73% CO_2 emission reduction within 30 to 35 years is considered. The growth of renewable energy during the last decades was sufficient for the energy transition if policies overcome the complementarity of energy resources. For example, 5–6% substitution of fossil fuels for renewable energy per year during 30 years can meet the growing demands for energy consumption along with the CO_2 emission reduction by nearly 80% (e.g., $0.95^{30} \approx 0.21$), which can attain CO_2 emission levels that are below those of the year 1990. The saturating growth of fossil fuels and accelerating growth of renewable energy during the late 1900s are positive indicators of this perspective. For a realisation of this perspective, the mechanisms that drive complementation and deforestation must be eliminated through concerted policy action in the coming decades. For this purpose, stakeholders and policies can create conditions of high risk for the activities in fossil fuel and low risk for renewable energy. If such conditions are created, fast mitigation of climate change can be expected along with larger accessibility of energy without endangering income.

A major issue is the skewed distribution of CO_2 emissions across countries because only a few countries have a large share in global emission. By 1990, three high-income countries – the USA, Japan and, EU – caused nearly 50% of all CO_2 emissions. Their share declined to less than one-third in 2015. The Russian emissions were also large and also declined, whilst the Chinese and Indian levels grew fast. Those six countries covered nearly two-third of the global CO_2 emission; the remaining one-third is divided between more than a hundred countries, each one with a small share. If these six countries move fast toward the substitution of fossil fuel technologies for renewable energy ones, the economic benefits of selling modern technologies go along with progress in the mitigation of further climate change. While climate change negotiations involve all countries, those six determine the global outcome for the climate.

During the last decades of the 1990s and early 2000s, CO_2 emissions grew steadily, but can turn around into a fast decline if decision-makers generate continuous substitution of fossil fuels for renewable energy. Even if this substitution rate is 5–6% a year, mitigation of climate change is in sight. Therefore, policy agreement between the six largest polluting countries is instrumental.

2.6 Conclusions

The questions of whether there are sufficient energy resources for manifold larger than the present scale of energy consumption in the future and if pollution can be sufficiently reduced through the shifts from fossil fuels to renewable energy are addressed. They are answered largely based on the available assessments of flows of renewable energy and reserves of fossil fuels in comparison with the interpretation of statistical data on energy consumption and CO_2 emissions. The short answer is that the renewable energy flows are plentiful whilst the fossil fuels stocks are scarce, aside from abundant coal, and they are polluting. A longer answer below explains this in more detail.

If only renewable energy would be consumed, global energy consumption could grow nearly infinitely because the flows of renewable energy on Earth are sufficient for a few thousand times larger energy consumption than what it was in the year 2015. Moreover, these flows are available at almost every place on the globe. However, these energy resources are low in density, which requires space; they vary by location and time which needs storage to overcome temporary shortages, and they cannot be easily switched on and off which impedes dispatching of power and heat unless supplied in networks. Even if dense energy stocks cannot be missed, nearly all high-density energy can be based on biofuels. Contrary to widely distributed, fluctuating flows, the stocks of fossil fuels on the Earth are concentrated in large quantities in the specific locations. These stocks are sufficient only for a few hundred years of the present energy consumption if the hypothesised technologies are taken into account. This is with the exception of coal whose stocks are larger than all other fossil fuels put together, which are sufficient for a few thousand years of present energy consumption.

The problem is not the scarce availability of energy resources, but the growing consumption of fossil fuel from the mid-1800s on when the global energy consumption increased about 20 times toward 2000. Meanwhile, renewable energy grew slower and its share declined from 98% in the year 1800 to 14% in 2000 whilst the remaining part is covered by fossil fuels. During these two centuries of growing energy consumption, the shift from traditional renewable energy mainly based on labour, biomass, and hydropower to fossil fuels evolved within 50 years, from the late 1800s to the early 1900s when fossil fuels expanded from a few percent in global energy consumption to more than half of it. In the late 1900s, modern renewable energy emerged based on geothermal, wind, solar resources, and biofuels which grew fast alongside the saturating growth of fossil fuels and traditional renewable energy. The economic booms in the late 1800s and 1900s enhanced these shifts.

In a result, nearly all energy resources grew throughout two centuries. Novel ones grew faster, thereby complemented rather than substituted resources vested in the past. This complementation mechanism can be explained by the combined effect of resistance to risky changes of energy resources by suppliers, and inventiveness of incumbents in finding profitable opportunities in market niches. In effect, specialisations in various markets emerge which diversify energy resources. This process goes on. From the late 1900s on, modern renewable energy expanded, which complemented rather than substituted fossil fuels. If this complementation mechanism continues along with growing energy consumption shortages of fossil fuels are faced within several generations unless major technological break throughs in fossil fuels are attained.

A more pressing issue is pollution caused by the combustion of fossil fuels. That shift from traditional renewable energy to fossil fuels caused 750 times higher CO_2 emissions in 2000 compared to the year 1800. Larger CO_2 emissions along with the decline of global biomass caused saturation of natural sinks for CO_2, followed by climate change that threatens economies on large parts of the Earth. While the combustion of fossil fuels causes CO_2 emissions, renewable energy resources do not emit CO_2 in consumption; an exception is a biomass that absorbs more CO_2 in the living plants and stores carbon after their death, which is released during combustion. In effect, the impacts on climate change, measured in CO_2 emission equivalents throughout the life cycle of renewable energy, are 20–100 times lower compared to fossil fuels, except nuclear power that causes similar CO_2 emissions per energy unit.

Regarding the saturating CO_2 sinks whilst growing energy consumption during the next decades, fast CO_2 emission reduction is necessary. For mitigation of further climate change, a minimum 73% emission reduction by 2050 compared to 2015 is needed. Therefore, the substitution of fossil fuels for renewable energy is required, which is possible when a 5–6% substitution per year is attained as a continuous trend. The main societal challenge is that the complementation mechanism impedes further CO_2 emission reduction. Policies can overcome this mechanism if the investments in fossil fuels pose high risks compared to ones in renewable energy which can be attained when all policy support for fossil fuels is withdrawn in favour of renewable energy and afforestation. Such policies obstruct the production of fossil fuels and generate budgets that can be allocated in socially more beneficial ways.

Chapter 3
Changing Energy in Economies

The question of what economic mechanisms have driven changes in energy consumption during the last two centuries is addressed using statistical data. It is argued that a larger population, due to higher income, triggered a higher consumption of energy and fossil fuels. As the growth of agriculture and industries during the 1900s declined while services grew, the growth of that consumption was also saturated and specialisation in energy services emerged. Given those changes in the economic structure, the changes in energy consumption are explained with reference to the neo-classical train of thought about prices of scarce energy resources, the evolutionary viewpoint focused on progress in energy efficiency of technologies, and the behavioural one about value addition by energy services.

3.1 Introduction

Economic interests that generated shifts in energy consumption from the traditional renewable energy to fossil fuels throughout the last centuries are addressed based on the statistical data. These observations of the past help to indicate possibilities in the future. Reference is made to robust changes in population, income, energy, sectors, and technologies during several decades or longer, across continents and countries. The continents are defined by the geography of present countries. North America covers Canada, Greenland and the USA. Europe covers all countries in mid- and west Europe. The former Soviet Union is separated; labelled in many databases as 'CIS' after Confederation of Independent States whereas most data refer to Russia. Oceania covers Australia, New Zealand, and the Pacific. The whole of Africa is considered. Asia includes Japan, Middle East and Turkey. Latin America includes Mexico and the Caribbean.

Y. Krozer, *Economics of Renewable Energy*,
https://doi.org/10.1007/978-3-030-90804-1_3

While the global population and income grew throughout the last two centuries, those growth rates varied across continents and over time. The economies grew fast in North America, Europe, and Oceania, often labelled as the Western countries or the 'West'. The CIS economy fluctuated. Meanwhile, the economies stagnated or grew slowly in Africa, Asia and Latin America, labelled as the 'South'. Accelerated growth of population, income and energy consumption on all continents is observed after the 2nd World War (1939–1945) when independent countries of the South are vested after the decolonisation, and economies of the West boomed. The drivers of faster growth in the West when compared to the South throughout the two centuries are much discussed. A popular political explanation refers to the Western military power which enabled the exploitation of cheap mineral and forced labour in Asia, Africa and Latin America for Western interests (Wallerstein, 1974). The environmental argument is that warm climates in the South were suitable for agriculture, whereas temperate climates in the West were better for industries that generated higher value (Lander, 1998). The sociologist argument is that elites in the South forced cheap labour of animals and people – especially women – which obstructed coal production, steam machines and other productive technologies (Pommeranz, 2000). These theories are not disputed. Herewith, the economic narrative about innovation is used not because it is most convincing per se on all continents, but because it most relevant for the assessment of energy production and consumption.

That innovation narrative pinpoints the origin of Western economic growth in the period of major socio-cultural changes in Europe during the 1700s. It was a period of political relaxation in the West after religious conflicts, epidemics, and wars that decimated its population from 1500s until the late 1600s. During this political relaxation in the 1700s called 'Enlightenment', liberties were demanded by the emerging social class of urban citizens. This middle class – or bourgeoisie – revolutionised France, took power in England, liberated North America from English and French colonialism, and pushed for abolishing slavery and serfdom, though they remained in practice. Economic relations based on the merits of individual capabilities in arts, knowledge, and politics were pursued. The primacy of individual capabilities undermined the principle of inheritance vested by the nobility and church for centuries. Many inherited entitlements and family patronages, however, remained; for example, in land ownership.

The economic importance of that liberal principle was that the ownership of technical discoveries and crafts could be arranged based on individual merits, which was defined in patents and copyrights across Europe. This way, large capital for risky investments in novel ventures could be attracted because the private property of mercantile results was assured, not only by inheritance and trade as it was formerly arranged through the entitlements of authorities. Along with business associations, trade houses, patent offices and other institutions that protected the interest of private investors, workers were freed to own and sell its labour which invoked pressures for higher incomes on the labour markets rather than bondage to land as slaves and serfs. Higher incomes generated demands for goods with mass production. Mortality decreased as people could pay for better food, heating with coal, lighting with whale oil, and other means for private consumption. Furthermore,

more people could benefit from the sanitation of sewerage, city lights with coal gas and other public services made available in cities. Lower mortality generated a larger, younger population, thereby economic growth.

In this narrative, the drivers of economic growth are considered entrepreneurs that seek fortune through discoveries of new resources, and the inventions of products perceived valuable by customers, called 'innovators' (Rosenberg & Birdzell, 1986). That innovation perspective helps to assess mechanisms of change in energy consumption in the past and possible changes in the future. Pinpointing at the societal, communities and entrepreneurial interests in renewable energy supports the formulation of incentives for changes, rather than authoritarian dictates which usually cause social discontent and economic stagnation after some time. For example, a study across nearly a hundred countries under authoritarian ruling shows an average 1.8% annual decline in real income after 10–12 years (Papaioannou & Luiten van Zanten, 2015). The authoritarian ruling also impedes innovations in energy (Rosenberg & Tarasenko, 2020), though links between democracies and energy are only indirect (Kammerlander & Schultze, 2020). After presentation of global changes in income, energy, and sectors, development in energy production and consumption are analysed from the neoclassic, evolutionary and behavioural economic perspectives.

3.2 Income and Energy Growth

The growth of global population, energy consumption, and income throughout the last two centuries are indexed with OWD data in Fig. 3.1 starting from 1800 (1800 = 100). Herewith, the population is measured by the number of people, energy consumption in TWh, and income in USD_{2011} based on GDP; note that much of the data is not available in databases and interpolated. Results are shown in 10 years intervals with many interpolations, in particular data for the 1800s are scarce. Given imperfect data, trends are relevant rather than being numbers from a particular year.

Until the 1800s, the global population remained below one billion people but thereafter, it grew to about 6 billion from the year 1800 to 2000 due to lower mortality in the West. It shot to 7.8 billion in the year 2020 due to lower mortality in the South. Despite large population numbers, the growth rate of the global population was only 0.6% from 1800 to the 1950s, which accelerated to 1.7% thereafter due to the improvements in the South; these growth rates declined after the year 2000. Meanwhile, global energy consumption grew faster than the population. The growth rate was 1.5% a year from 5.6 PWh in 1800 to 112 PWh in 2000, which was 0.9% during the 1800s and 2.1% over the next 100 years. Global income grew even faster. This growth rate was 2.1% annual average from USD_{2011} 842 billion in the year 1800 to nearly USD_{2011} 63,100 billion in 2000. The growth rates fluctuated. While the global income grew by nearly 1.0% from the year 1800 to 1850, it accelerated in the late 1800s in the West, slowed down in the first half of the 1900s because of wars and economic depression of the 1930s, and accelerated to

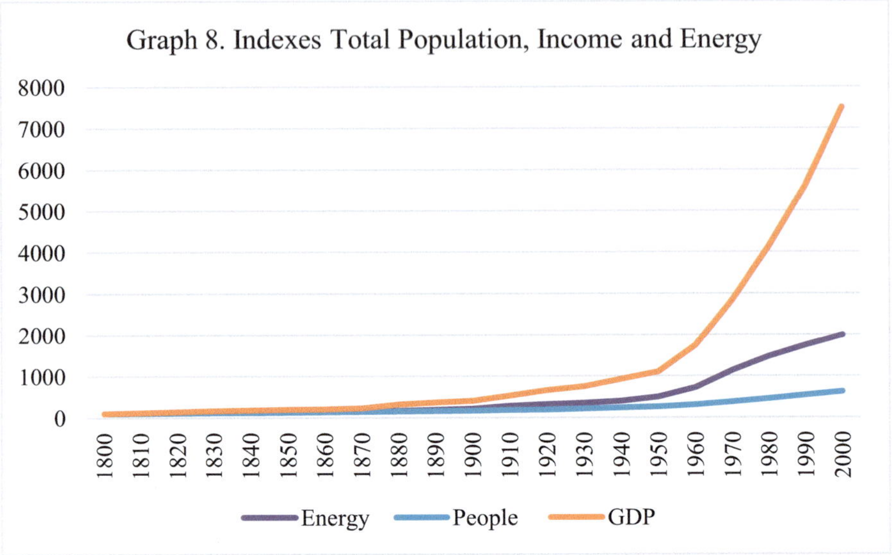

Fig. 3.1 Indexes total population, income and energy

Table 3.1 Growth of income in USD_{2011} per capita across the continents based on the Maddison database

GDP growth per capita	North America	Europe	Oceania	Russia (CIS)	Africa	Asia	Latin America	World
1800–1850	0.8%	0.5%	3.1%	0.03%	0.3%	0.2%	−0.03%	0.4%
1860–1900	1.6%	1.4%	1.9%	0.7%	1.3%	0.5%	1.1%	1.2%
1910–1950	1.8%	0.9%	1.7%	2.2%	0.9%	0.3%	1.6%	1.3%
1960–2000	2.2%	3.2%	1.7%	1.2%	1.4%	2.8%	2.2%	2.2%
1800–2000	**1.6%**	**1.5%**	**2.1%**	**1.0%**	**1.0%**	**1.0%**	**1.2%**	**1.3%**

3.7% in the second half of 1900s when economies in the South grew faster than in the West. During the 1900s, the global population grew slower than energy consumption, which grew slower than income. While the global average income per capita increased 12 times, energy consumption per capita increased tripled. The world became more populous, energy better accessible and people richer during the last two centuries.

The growth of income per capita varied across continents and over time as shown in Table 3.1, based on the Maddison database. However, these differences have declined during the 2nd half of the 1900s, meaning that the incomes per capita have converged across continents. For example, while the population of Asia and Europe together covered 91% of the global population in the year 1800 it declined to 74% in 2000, and their incomes declined from 85% to 54% of the global income.

Throughout these two centuries, the growth rates of income per capita (bold) were higher in the West than in the South. Meanwhile, the growth rates steadily increased in North America, declined in Oceania, and fluctuated overtime on other continents. The fluctuation patterns in the growth rates differed as the African growth rates followed the European ones, the Latin America growth rates the North American ones which indicate the economic links, whereas the Asian income per capita grew fast only in the late 1900s. While the global average income per person grew more than tenfold from nearly USD_{2011}1000 in 1800 to nearly USD_{2011} 11,000 in 2000, major changes occurred across countries. For example, an average European was poorer than a Russian in the year 1800, and twice as richer in 2000, both were colonial powers. Meanwhile, North American and Oceanian immigrants who were the poorest in 1800 across the continents became the richest in the year 2000. For example, a typical Russian with USD_{2011} 1680 in the year 1800 was more than three times richer than an Oceanian immigrant, but that immigrant became three times richer 200 years later. An average Indian, Egyptian, Syrian, or Jamaican was richer in 1800 than a Japanese or German, but poorer 200 years later. Within Europe, the income differences remained; for instance, a Swiss remained about six times richer than a Romanian. While people in most Western countries became steadily richer, the average income per person heavily fluctuated during two centuries, particularly in Russia, China and Vietnam in Asia, Egypt and Tunisia in Africa, and Venezuela and Uruguay in Latin America. The growth factors and distribution of income are beyond the scope of this book, and comprehensively covered in other works (e.g., Milanovic, 2012; Piketty, 2014).

Unfortunately, the data about energy consumption in countries and on continents during the 1800s are deficient, and qualities dubious. For example, the HYDE database provides such data with nil energy consumption in Africa and Asia in the early 1800s, which is implausible. Therefore, the last century is covered by combination of production data from the SHIFT database for 1900–1920 with consumption data from UN statistics for 1930–1950, and OWD data for 1960–2000. This production data is used assuming that energy production did differ from consumption across countries in the early 1900s, not across continents because the intercontinental transport of fuels expanded in the second half of the twentieth century; oil tankers started to operate in the late 1800s. Given the data imperfections, only the trends are relevant. Those data indicate that energy consumption per capita differed across the continents and in time. The volumes per capita in North America and Oceania were higher than in Europe, whose energy consumption in the early 1900s was higher than in Russia and the South. However, the growth of energy consumption per income in North America, Oceania, and Europe was slower than in Russia until the 1980s when its economy collapsed, and was slower than in Africa, Asia and Latin America where these growth rates were threefold higher. The data indicate that Russia and many countries in the South experienced more energy-intensive economic growth than the West.

Table 3.2 Annual average growth of per capita income, energy consumption and fossil fuels during 1900–2000 based on the Maddison database

1900–2000	North America	Europe	Oceania	Russia (CIS)	Africa	Asia	Latin America	Total
Income	2.0%	2.1%	1.7%	1.7%	1.1%	1.6%	1.9%	1.7%
Energy	1.4%	1.6%	2.0%	3.1%	4.9%	4.7%	5.8%	1.8%
Fossil fuels	1.2%	0.9%	2.7%	3.4%	6.0%	4.2%	5.8%	1.6%
Correlations income-energy R^2	0.00	0.55	0.04	0.88	−0.66	0.29	0.06	0.45
Correlation income-fuels R^2	0.28	0.63	0.49	0.84	−0.18	0.24	−0.15	0.42

Those observations are relevant with regard to the debate about the energy-driven of economic growth. Scholarly opinions about the relation between the countries' income and energy growth differ. A meta-review of 686 studies shows four modes of thought that reflect various assessment methods: no direct relation is observed; income generates energy, energy growth generates income, and finally feedback loops are pinpointed (Kalimeris et al., 2014). From the environmentalist perspective, it is often argued that higher income is driven by larger energy consumption. In particular, larger consumption of fossil fuels would explain the income growth during the 1900s because the higher energy densities of fossil fuels over those of traditional renewable energy would enable larger power. This argument is underpinned with correlations between income, consumed energy, and fossil fuels; for example, the historical correlations for several countries (Ayres & Voudouris, 2014) and across countries (Beaudreau & Lightfoot, 2015). Based on that argument, the decreasing income is advocated for CO_2 emission reduction, because such degrowth of economies would reduce consumption of energy, and fossil fuels. However, the growth of income and energy consumption poorly correlated over time. Herewith, the annual growth rates of per capita income, energy production, and fossil fuels on the continents are correlated for the 1900s. The growth rates of income, energy consumption, and the correlations are shown in Table 3.2.

Per capita, the incomes grew faster than energy consumption, and fossil fuels in North America and Europe, nearly as fast in Oceania, while slower in Russia, Africa, Asia, and Latin America. Fossil fuels consumption per capita grew faster than energy consumption in Oceania, Russia, Africa, and the growth was similar in Latin America. The economic development becomes more energy-intensive in the South where it was also more fuel-intensive except in Asia. However, the high correlation between the growth of income and energy consumption is found only for Russia, which is moderate for Europe, and low for all other continents, even negative for Africa. The correlation between income growth and fossil fuels is also high for Russia, but it is moderate for Europe and Oceania, and low or negative for other continents. These observations indicate that another factor than income growth drives consumption of energy and fossil fuels. This finding is robust because low correlations are also found for the annual growth of per capita income in USD_{2011} and energy consumption in cases of 14 most populous countries introduced in Chap. 1, as well as across all EU countries during 1990–2015. Higher regressions that 0.5 ($R^2 > 0.5$) are found only for Russia, India and, Nigeria, as well as only for Sweden within the EU. Thus, various economic patterns are more important for energy consumption than income growth; nevertheless, the idea of close links between income and energy consumption persists in studies on the history of energy systems (Smil, 2000).

Low correlations should not come as a surprise when the economic viewpoint is taken. Usually, the costs of energy resources are small fractions of the costs per unit energy consumption because all conversions of energy resources involve mainly capital and labour. Furthermore, the energy services as result of these conversions rarely cover more than 10% of the incomes per capita on all continents, excluding subsidies for energy. When these two factors are combined, the costs of energy resources are rarely above a few percent of income even in 'industrialised countries', 'manufacturing hubs' and suchlike labels for the energy-intensive businesses. It indicates that the costs of energy resources are rarely essential for income growth; more important is the temporary impact of high energy prices on inflation because it takes time to counter higher prices through energy saving. Although the share of energy producers is a maximum of a few percent of the national income the idea that they are drivers of the income growth is persistent and widespread. Apparently, their influence on ideas and policies is disproportionally larger than their share in income; high concentration of capital in these industries could be a reason for this impact. A major weakness in the analyses of the national income and energy is high aggregation because the sectoral performances are important. For instance, economies based on basic industries use necessarily more energy than ones based on agriculture, and firms in the basic industries with obsolete technologies use more energy than ones with state-of-art technologies, whereas obsolete production continues when authorities protect them from innovative rivals or entitle for a monopoly position through permits, royalties or certificates. Under competition, changes in the economic structure and performance are decisive for the growth of income compared to energy consumption.

Different patterns in economic development can be explained by changes toward more valuable production due to growing services as higher value is generated when materials, energy, capital and labour are combined by knowledge into qualities that are highly appreciated, and paid for by the consumer. The global value added grew during the 1900s, which means that the resources of capital, labour, and knowledge added more value to the purchases of energy and materials in constant prices. The value-added grew in agriculture and industries while the highest growth rates on all continents were in the services, particularly fast in North America, Europe and Oceania. Figure 3.2 shows the shares of value-added in the agriculture, industries and services on continents in the years 1905, 1945, and 1995 based on the HYDE data. This data covers the period from 1895 to 1995 in five-year intervals; the earliest intervals are excluded because of dubious data.

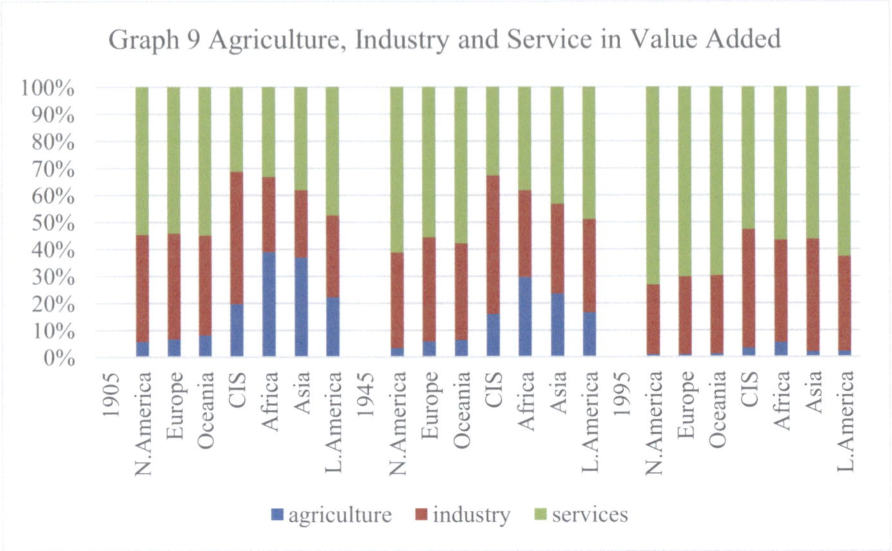

Fig. 3.2 Agriculture, industry and service in value added

As heavy industries need energy-dense resources for cheaper production of metals, building materials, glass, and other bulky products, these growing industries during the 1800s have generated the shift from biomass – available on farms – to the energy-denser fossil fuels (Smil, 2004). Hundred years later, the share of service in the global value increased yearly by 0.2% in the 1900s, whilst the share of industries declined yearly by −0.1% a year and agriculture by −0.9% a year. In effect, the share of services in the global economy increased by 26% in 100 years. Different patterns across the continents are observed. In 1905, the share of agriculture together with industries was about 40% in the West, nearly 70% in Russia (CIS); and 50% to 60% in the South. While the share of services grew in the West to 60% in 1945 and exceeded 70% in 1995, it grew slower in the South and even slower in Russia (CIS) where only 50% of services in the economy was attained in the year 1995.

A popular argument about the growing share of services in the West is so-called outsourcing or out-shoring, which means a shift of the production capacities in the manufacturing activities from the West to the South. Though the growth of manufacturing in low-income countries is indisputably observed this argument does not fully explain the growth of value-adding services. The growing share of services in economies is a trend observed on all continents and the growth of value-added in economies is closely correlated to the services on all continents ($R^2 > 0.9$ based on HYDE data), and the growth of services is less energy-intensive than the growth of industries. However, the growth of services is not the only factor in the explanation of changes in energy consumption regarding the high growth of energy consumption compared to income growth in Asia and Latin America along with a higher share of services in their economies. Presumably, policies and cultural factors in energy consumption are also relevant; for example, much energy is wasted when it is cheap because subsidised.

Table 3.3 Installed energy capacity in USA from 1850 to 2000, based on Grübler (2012)

Sectors in energy consumption divided into the industries, I, and consumer-based services S		Installed capacity of energy in GW 1850	1900	1950	2000	Growth per year
Stationary End-users	Thermal: furnace, I	300	900	1900	2700	1.5%
	Mechanical: movers, I	1	10	70	300	4.0%
	Electric: appliances, S	0	20	200	2200	4.9%[a]
Mobile End-users	Non-automobile, S	5	30	120	260	2.7%
	Automobiles, S	0	0	3300	25,000	4.2%[b]
Stationary Suppliers	Electric power, I	0	10	260	2600	5.8%[a]
	Mechanical movers, I	0	3	70	800	5.9%[a]
	Chemical refineries, I	0	8	520	1280	5.4%[a]
Sub-totals	Industries	301	931	2820	7680	2.4%
	Services	5	50	3620	27,460	4.2%

[a]Average of 1900s [b]1950–2000

Along with all services, the energy services expanded and diversified from the early 1900s on because tuned to the consumers' demands for services. The growth and diversification of energy services are illustrated in Table 3.3 with the data on installed energy capacities across sectors in the USA, from 1850 to 2000 in intervals of 50 years (Grübler, 2012).

While total installed energy capacity in the USA increased nearly 36 times from the year 1850 to 2000, that total increase was generated mainly by the consumers' services. By 1850, nearly all energy capacity was installed in industries, whereas by 2000 about 78% of all was generated by the consumers' services, mainly in automobiles. The remaining 22% was mainly covered by the specialised supplies, in particular the electricity services. Given the high share of services in the energy capacities, it is not surprising that the use of automobiles in renewable energy networks is pursued; for example, storage of generated electricity in batteries. While this data refers to the USA with high car mobility, similar changes evolve across countries.

The growing service also influenced CO_2 emissions. Figure 3.3 shows indices of global CO_2 emissions in million tonnes, as well as tonnes CO_2 per capita, per kWh and per USD_{2009} from 1800 to 2000 (1800 = 100), all based on the OWD data.

As the global population and energy consumption increased during the last two centuries, total CO_2 emissions also increased exponentially. These emissions per capita, per energy unit, and per USD_{2009} also increased during the early phase of economic development until the late 1800s in a linear fashion; however, their growth saturated during the 1900s and even decreased after the 1980s. This saturation is driven by a larger share of services, as it is less energy-intensive than industry.

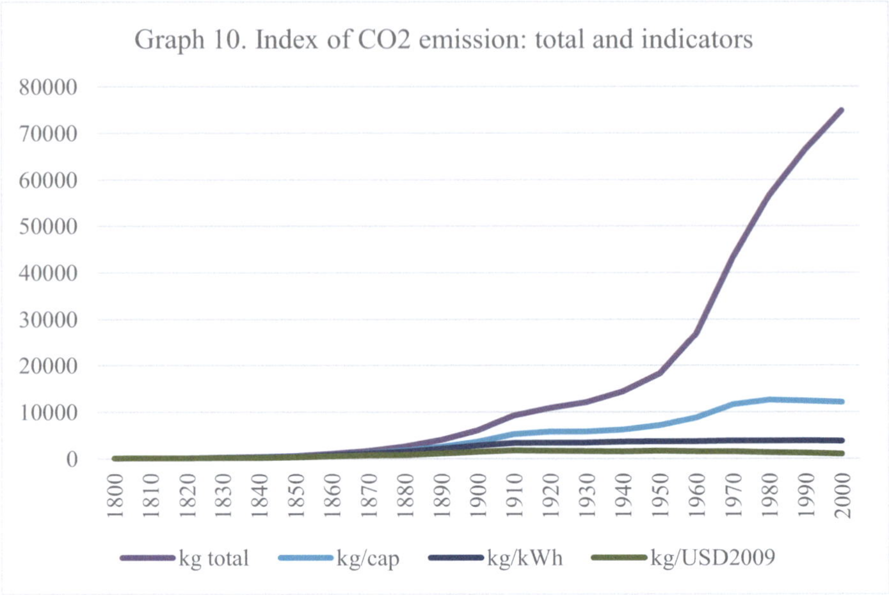

Fig. 3.3 Index of CO2 emission: total and indicators

Moreover, CO_2 per unit energy consumption hardly increased after 1920, even decreased from the 1980s onwards due to the larger share of services in economies, as well as the growing share of fossil fuels with lower carbon content, and the growth of modern renewable energy.

The global population grew during the last two centuries due to lower mortality and better livelihood when industries generated higher incomes on all continents. These were the main drivers of larger energy consumption and CO_2 emissions whilst growing services invoked decoupling of income per capita from energy consumption and triggered energy services tuned to consumer demands. Below more specific drivers for growing fossil fuels during the late 1800s are discussed based on the neoclassical, evolutionary and behavioural theories.

3.3 Energy Prices and Consumption

Mainstream, neo-classical economics is focused on prices in the transactions between competing suppliers and demanding consumers. An interplay between the energy supply prices and the scale of consumption is assumed. When scarcer energy resources cause higher supply prices, substitutions for cheaper energy resources are invoked followed by lower prices which trigger larger energy consumption of these cheap resources whilst larger consumption invokes higher prices, and so forth. This price mechanism is observed from before the 1800s. When forests in Western

Europe were largely cut for agriculture and timber, the real prices of fuelwood increased sixfold from the early 1500s to the mid-1700s because of costly imports of wood. During several centuries, the prices of energy resources in the West grew faster than the prices of agricultural and industrial products (Taylor, 2017).

The higher wood price along with larger energy consumption could have triggered coal production, and higher coal prices during the 1700s and early 1800s could have invoked oil, but there are also doubts. It is argued that industries kept using wood whilst households used coal in England in the 1800s – this country was the main producer of coal at the time. Moreover, coal-based steam machines were not the drivers of income growth in England. This is argued regarding the slow productivity growth in steel, clay, and other energy-intensive industries. Meanwhile, fast increasing productivity in textiles and other consumer-oriented industries used less energy per product (Clark & Jacks, 2007). In the USA, which was the largest coal producer during the 1900s, wood was also widely used in industries along with coal while the coal prices hardly influenced the uses of wood and hardly increased the industrial energy efficiency. The industrial energy efficiency remained low because wood and coal were abundant, and taxed low for applications in businesses (Rosenberg, 1994). It means that the energy prices and consumption were not as closely connected as the mainstream theory suggests.

The main shift in energy resources was the expansion of mineral oil for energy consumption from the mid-1800s onward, though small scale production emerged early 1800s. As global mobility grew about 1.4 times faster than the income throughout the 1900s, oil production for the auto-mobility grew to become the largest energy resource and the benchmark for energy prices. Oil is mainly used in mobility while coal remained the main resource for stationary combustion. Below, the interplay between oil prices and demands is assessed in some detail because illustrates several factors that influence prices of energy resources. The prices in the current dollars and constant dollars of 2018 between the years 1861 and 2017 are shown in Fig. 3.4 in USD_{2018} per barrel oil equivalent (b.o.e.) in the USA, based on the BP data; recall one b.o.e is equivalent of 0.146 tonnes oil equivalent (t.o.e).

For nearly 30 years after the start of large-scale oil production in 1861 in the USA, its annual average production grew by 15% while real oil prices decreased by nearly 10% along with price swings from USD_{2018} 40 to USD_{2018} 120 per b.o.e. Given the growing demands in consumption, the decreasing prices can be explained by the decreasing costs per unit production due to larger scale of production and the swings by the temporary oil scarcities followed by a larger scale of production at the lower unit costs, thereby the decreasing oil prices. Those changes are in line with the neo-classic theory. The demands for oil were enhanced by taxes put on the rival biomass-based alcohols, which were the main fuels in the USA until the early 1900s. Although comparisons of the taxes on bio-alcohols to their market prices of that time were not found, these taxes were high compared to the prices as they covered the largest part of federal tax income early in the 1900s (Encyclopedia.com, 2020). So, the emergence of oil production in the mid-1800s can partly be explained by the decreasing market prices of oil compared to the prices of biomass and coal as the theory predicts, and partly by the state support of oil consumption through the taxes on the rival bio-alcohols.

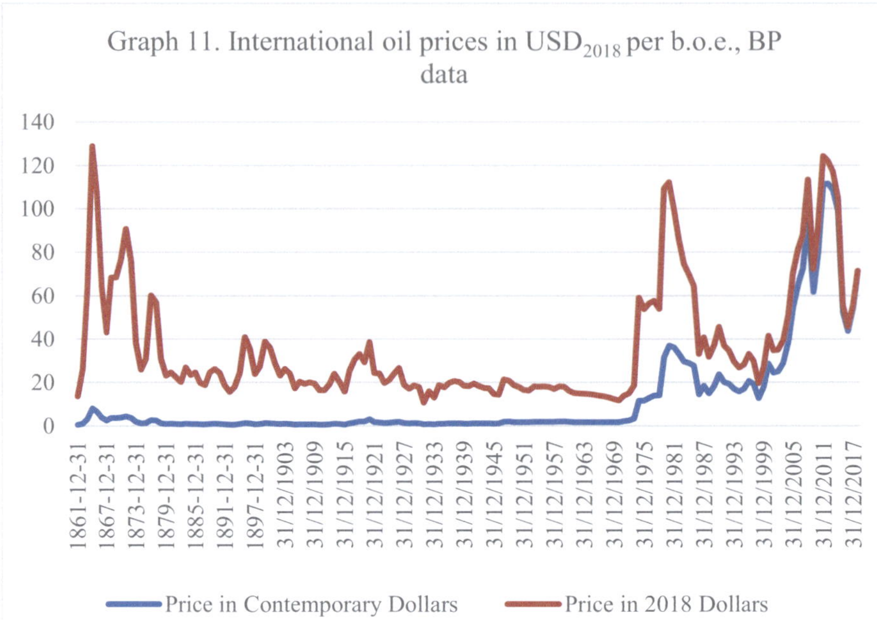

Fig. 3.4 International oil prices in USD_{2018} per b.o.e., BP data

During the subsequent 70 years between the early 1900s and 1970s, global oil consumption enlarged nearly 150 times, at a 7% annual average growth. Meanwhile, the oil price hardly decreased and mildly fluctuated around USD_{2018} 20–40 per b.o.e a year; they decreased throughout those 70 years only by −0.2% annual average. Based on the mainstream theory, the decreasing unit costs of a larger production scale can be expected, whereas steady prices rather than decreasing ones are often explained by monopolies that kept prices above the competition, so as to have higher profits. Whether this explanation holds true is subject to debate that is not discussed. Only the idea about 'natural monopoly' is introduced because emerges time and again in discussions about private and public energy services.

The idea of 'natural monopoly' – situations when a firm can avoid competition and controls essential resources – is justified by the necessity of large investments for generating of large-scale operations. This was facilitated by the regulation that allowed to shift the liability from the capital owners to a chartered corporation, a holding with a number of trustees, thereby reducing the risk borne by investors. This regulation triggered a financial conglomerate that monopolised the oil production, the Standard Oil Corporation in 1870 in the USA. While monopolies were prohibited, this one was justified with the argument that large investments in the expansion of the upstream oil production in mining with downstream refining and sales need a conglomeration of finances. By 1911, Standard Oil was broken into several companies by the decision of the Supreme court. However, oligopolies – a few competitors in a market – were acceptable in the USA, while monopolies and price agreements were tolerated in Europe. For example, monopolies were justified in England with reference to entitlements for exploitation of land given by royals to owners in the 1600s, and the price agreements between producers of lamps made in the early-1900s were tolerated for 80 years. After the 2nd World War from 1939 to 1945, monopolies were tolerated in the USA and Europe on the argument that the economies of scale are efficient on the markets of natural resources. However, this argument about the 'natural monopolies proved to be a hoax exploited for the protection of public utilities in energy during the 1950s, and privatisation of them during 1980s (Melsted & Palua, 2018). Moreover, once privately owned upstream production of oil became state-owned monopolists in many countries during the late 1900s, yet downstream refining and sales remained private. This illustrates that the oil prices were also influenced by the state interventions.

From 1971 to 2018, oil production grew at an annual average of 3%, whilst the prices increased 9% a year due to two price peaks. During those peaks, the prices rose from USD_{2018} 20 per b.o.e. to USD_{2018} 120 per b.o.e. The first price peak between 1974 and 1985 was explained by the agreement about production limits by the oil producing countries in the Organisation of Petroleum Exporting Countries - OPEC. As a side-effect of this 'oil shock', the biomass being processed into bioethanol, biodiesel and biogas grew, and wind energy and solar energy were introduced. After 1984, the prices dropped to the levels before the oil shock, only to peak 20 years later while that modern renewable energy kept growing. Next, the period of high oil prices from 2005 to 2015 is explained by the growing demand for energy in populous Asian countries as a result of fast income growth. This 'demand shock' enhanced the production of modern renewable energy and triggered the introduction of modern renewable energy with energy storage with batteries in distributed energy systems without grid. Fossil fuels based on shale oil, tar sand oil, and gas fracking also expanded in North America; they refer to rocks with petroleum-like fluids (kerogen), to sands with heavy hydrocarbons, and to pressing gas with water and chemicals, respectively. However, these price agreements were confined by the states that pursued larger production. For instance, Venezuela expanded its oil production in 1985 and the USA opposed agreements between the oil producers in the 2000s when it expanded oil production in tar sands and oil shale (Bunger, 2011). Hence, the oil prices were mainly influenced by state interventions, whilst modern renewable energy grew during low prices between the shocks, when oil prices dropped to USD_{2018} 20–40 per b.o.e.

High, swinging oil prices from the 1980s on are interpreted by many environmental experts as signals of emerging resource scarcities. Those swings are assumed to reflect the declining reserves at those locations where the exploitation of energy resources is cheap, whilst the exploitations of new sites are assumed costlier and mining results less certain. This so-called 'Peak Oil' theory disputed the sufficiency of oil for the twenty-first century (Ghosh & Prelas, 2009). However, subsequent assessments of the global reserve of fossil fuel during the 1900s showed an increase in fossil fuel reserves because exploration technologies improved (Shafiee & Topal, 2009). It is also underpinned that the real prices of fossil fuels decreased throughout the 1900s weighted for fuels shares in global energy consumption, which is attributed to progress in exploitation technologies (Shafiee & Topal, 2010). From the perspective of mainstream economics, the assessments of larger reserves due to better technologies are considered as worrisome misconceptions about the fuel scarcities, because the advances in the exploitation of energy resources cover up the overexploitation of fossil fuels (Nordhaus, 2010). Whoever is right about the scarcities, technologies in the production of fossil fuels and renewable energy progressed during low and high prices which underpin the idea about the quasi-autonomous process of generating know-how for cheaper technologies as is pinpointed in the evolutionary thoughts.

The mainstream idea that the higher prices of resources trigger a decline in energy consumption is also disputable because the link between energy prices and consumption is obscure when those price swings are considered. An increase in the unit price is usually assumed to reduce less than unit consumption because energy is a basic good for all kinds of consumptive activities, called low price elasticity of demand. A literature review of the studies on the price elasticity of energy demands across countries indicates that one unit price increase reduces 0.4 units of energy consumption in the short run, and 0.7 units in the long term, which are price elasticities of −0.4 and − 0.7 (Labandeira et al., 2015). Those studies on the price elasticity address the oil markets in the 1900s when the prices fluctuated moderately. If the whole period from the year 1900 to 2010 is considered based on the BP data for the prices and the mix of OWD and BP data for the consumption in 10-year averages, the changes are random, the estimates of price elasticities flip-flop from a positive to negative depending on the period taken into consideration. Also, nil correlation between the prices and production is found for the BP oil production and price during the period 1965–2018.

These results confirm the argument that the price fluctuations of natural resources are stochastic changes rather than reflections of scarcities (Ahrens & Sharma, 1997). It is also observed that the increasing supply prices do not necessarily reduce consumption, but rather increase it when innovations are generated (Arora, 2014); larger consumption of novelties can be caused by high consumers' appreciations, and prices increases when the state support of innovations creates monopolies, so-called 'ripple effects' of the state support. Furthermore, the complementation in energy consumption collides with the mainstream perspective because novelties rarely replace incumbent resources and technologies, entailing the diversification towards natural gas, hydropower, and nuclear power during low fuel prices in the 1900s, and modern renewable energy during high prices.

High prices invoked novel energy resources. Thereafter, the growth of energy resources is influenced by prices as well as the state interventions, market structures, changing demands, and other factors that are difficult to disentangle.

3.4 Energy-Efficient Technologies

From the evolutionary perspective, the technological change due to the generated practical knowledge from the past determines prices, which reverses the causality in the mainstream theory. That know-how is generated mainly through larger engineering capabilities fostered by education, research, and development which are largely supported by policies. Those engineering capabilities, in turn, enable the introduction and dissemination of novel technologies. In energy, this evolutionary argumentation about technologies focuses on a higher density of resources measured by energy content per mass. Higher densities of the energy resources are considered to be the key production factors as they enable higher power, and thereby lower costs to the economy (Hall, 2017). Therefore, low-density wood, peat, grass, and lignite (4–5 MWh per tonne) shifted during the 1800s to 1.5 times energy denser coal, during the 1900s to 1.6 times denser oil, and during the late-1900s, and early-2000s to 1.2 times denser natural gas. Along similar lines, hydrogen and uranium oxide are expected to expand because they are nearly three times and 4000 times denser than gas.

No doubt that the higher densities of those energy resources were relevant for the engineering of industrial processes, though not necessarily cheaper if the processes became less manageable and not always more effective for consumption if resources are wasted. Furthermore, performances of subsequent technologies in the value chains of production, distribution, consumption, and disposal – it is in the so-called life cycle – and during lifetime of technologies, are relevant for the cost of energy consumption. The life cycles of energy resources are relevant as some technologies require a lot of energy resources. For example, the production of hydrogen with the steam reforming of natural gas is exothermic; meaning that it uses more energy input per produced unit of hydrogen than the energy content of that hydrogen (Spath & Mann, 2001). For energy-efficient consumption, the sum of energy resources in the life cycle must be low. This is measured by delivery of energy services compared to the sum of energy resources used per step in the life cycle of that service, called the 'energy return on investment' (EROI).

Whether the energy-density of resources is decisive in consumption is uncertain because many processes are net consumers of energy whereas the EROI assessments for an exemplary energy service vary by a factor of 2 to 10 (Gupta & Hall, 2011), and even the EROI rankings of energy resources differ across studies (Hall et al., 2014). Comparisons of renewable energy to fossil fuels based on EROI assessments are also inconsistent when the estimates convert the stocks of fossil fuels into the flows of renewable energy with energy density per m^2 because such conversions depend on disputable assumptions about the fossil fuels reserves (Layton, 2008). When the EROI of consuming renewable energy is compared to fossil fuels, comparable functional qualities of energy systems must be defined; for example, large-scale thermal power plants versus solar roofs have different functionalities because continuous heat and electricity versus intermittent electricity, respectively (Cheng & Hammond, 2017). More consistent are simulations of the energy resources for specified production systems; for example, a comparison of fossil fuels and renewable energy for power generation in the United Kingdom

(Raugeri & Leccisi, 2016). However, the data quality and accounting methods in such assessments are also disputed.

Besides imperfect data and methods, the results of life cycle assessments depend on the normative choices of assessors, which are often implicit in estimates (Krozer & Vis, 1998). The main normative choices refer to the definition of system boundaries, inputs allocation, and unit of service. Three examples of biases illustrate these normative choices. Then first one is about the system boundaries for nuclear power. This energy resource is assumed to be of high density, but if all waste streams in the life cycle of a nuclear power plant are included within the system boundaries on the annual basis of that power plant, the EROI can be found low because a lot of energy is used for construction, dismantling and storage of all materials (Lenzen, 2008), and energy-dense waste is generated and stored for centuries (Eriksson, 2017). The second is about the input allocation for bioenergy. Although biomass is of a low energy density, the bio-residues for local services can have a high EROI because the bio-residues are nil energy inputs, contrary to the production of crops and wood that do need energy-intensive inputs (Narodoslawki, 2019). The third one is about the unit of service in hydrogen. The service of hydrogen produced with the thermal conversion of natural gas has a negative EROI when hydrogen is combusted for energy as envisioned in policies of many countries, whereas if it is applied for fertilisers in agriculture, it generates a positive EROI when agriculture produces high-calorific foods (Ramirez & Worrell, 2006).

Furthermore, changes in technology are particularly important for assessing the EROI. Higher EROI over time is often expressed as the effect-increasing technical change. Given that the data pertaining to the effect-increasing technical change in energy production is scarce, a few databases are combined for the assessment of changes in electricity production plants based on the combustion of coal. Figure 3.5

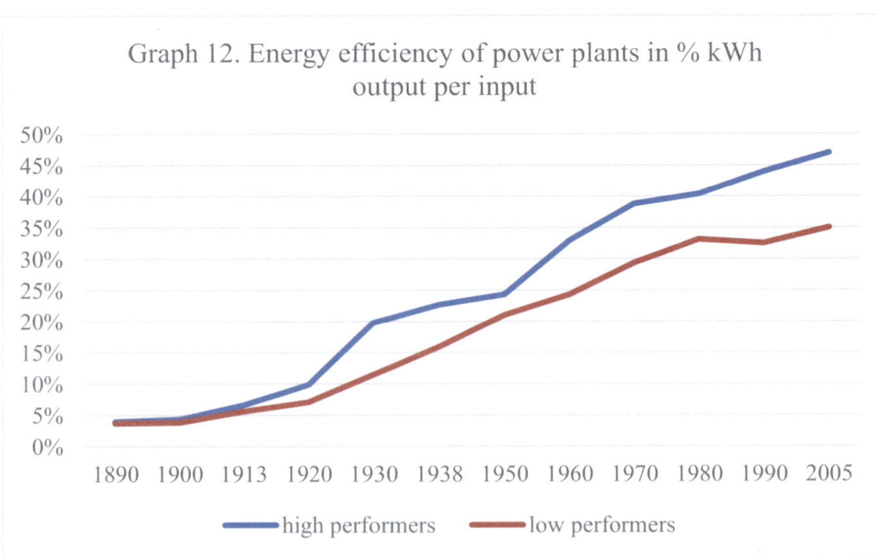

Fig. 3.5 Energy efficiency of power plants in % kWh output per input

shows the changes of high and low-performing plants from the year 1890 to 2005; note that the X-axis is not proportional to the timeline. The data on energy production from 1890 to 1938 in the Shift database are combined with the data of Etemad and Luciani (1991, p. XXXI–XXXV), which distinguishes high and low performers, as well as with the EU data from 1950 to 1980, and from 1990 to 2000 for high performers (EEA, 2011), and from 1990 to 2005 for low performers (JRC, 2012). All those data are converted into kWh; for example, tonnes of coal in kWh with the energy densities shown in Table 4.

The energy efficiency of electricity production was about 4% in 1890 which improved to nearly 50% among high performers and about 35% among low performers. While the annual average rates of effect-increasing technical change throughout the period from 1890 to 2005 were 4% for high performers and 3% for low performers, these rates declined over time. The rate of the high performers was 8% a year from 1890 to 1930, 3% during the next thirty years, but only 0.8% from the year 1970 to 2005. Further, this data indicates that the high-performers have attained an advantage in energy efficiency compared to low performers during the first forty years and that this advantage has remained during the subsequent seventy years. This so-called 'first-mover advantage' seems to be an important attribute in competition for energy-efficient power generation. Such first-mover advantages are also observed in other industries. For example, innovators in the chemicals kept markets leadership for decades because the innovators adapted their production, products, and marketing at a faster rate than the followers (Stobaugh, 1988). First-mover advantages can also be relevant for renewable energy but reliable empirical data is not found.

The declining energy efficiency over time is widely observed. For instance, the lumen per Watt increased 26 times between 1883 when the first Edison filament lamp was introduced and 1992 when the compact fluorescent bulb was introduced in the lighting market. This implies a 3% effect-increasing technical change a year throughout that period. The change from 1883 to 1920 was 4%, followed by a nil change because the lights business cartel impeded innovations until 1992 when the fluorescent bulb was introduced and performed twice better than conventional bulbs (Nordhaus, 1994). Thereafter, the efficiency-increase was only about 1% during 30 years compared to LED that disseminated from the early 2000s onwards, while the prices of LED decreased, and lifetime increased. In effect, costs per lumen decreased, measured throughout the life cycle on the annual basis which is so-called life cycle costs or total costs of ownership. Similar changes are observed in renewable energy. For example, the photovoltaic panels for electricity production (PV) captured 4% of solar irradiation in 1975, 25% in 2000, and even 48% in 2020 in the pilot plants. The annual average 6% effect-increasing technical change throughout nearly forty-five years was composed of 8% during the initial 25 years, and 3% thereafter (NREL, 2020). The effect-increasing technical change throughout many decades is essential for energy efficiency in the life cycles, whereas long-run changes can be overestimated based on the initial rate of change.

More effective conversions can reduce costs over time, called 'cost-reducing technical change'. Further, the cost-reducing technical change is also driven by greater know-how, a larger scale of application, and other factors not necessarily

Table 3.4 Rates of the cost-reducing technical change in energy based on Rubin et al. (2015)

Energy technologies	Unit costs decrease per year		Assessment period
	Minimum	Maximum	
Coal	5.6%	12%	1902–2006
Natural gas, gas turbine	10%	22%	1958–1990
Natural Gas Combine Cycle (NGGC)	−11%	34%	1980–1998
Nuclear power	negative	6%	1972–1996
Solar PV	10%	47%	1959–2011
Wind turbine on shore	−11%	32%	1979–2010
Wind turbines off shore	5%	19%	1985–2001
Biomass power generation	0%	24%	1976–2005
Biomass production	20%	45%	1971–2006
Hydroelectric	1.4%	1.4%	1980–2001
Oil Extraction (*)	5%	25%	1869–1997

(*) Estimates for Oil Industries (MacDonald & Schrattenholzer, 2003)

directly linked to more effective conversions. For instance, 4% effect-increasing technical change in coal power plants can largely explain the cost-reducing technical change in coal technologies. In PV, that 6% effect-increasing technical change is only a part of the two-digit cost-reducing technical change. Meanwhile, several studies have shown the minima and maxima of the cost-reducing change of energy technologies. Table 3.4 summarises results with the observation periods based on the literature review (Rubin et al., 2015). All rates are estimates of average annual changes; positive rates imply that costs are reduced, whereas negative ones mean that the costs increased.

While this assessment shows that unit costs of all renewable energy decline, a large spread in unit costs is also observed within the renewable energy technology and across these technologies. For example, the rates of cost reduction in wind power onshore vary from −11% to +32% a year in various studies. These cost-reducing technical changes also vary across technologies from 1.4% a year in hydropower, to 24% in bioenergy, and 47% in PV, which mean a decrease to half of the unit costs after 50 years, 5 years, and 2 years, respectively. However, the estimated rates can also be influenced by the length of observations because the assessment periods vary from more than one century for coal and oil, to a few decades for wind offshore and biomass (Yeh & Rubin, 2012). Therefore, policies aiming at the mitigation of climate change should consider the lowest costs CO_2 reduction of technologies at a particular moment, as well as the possibilities of progress. This can influence the policy support for particular renewable energy; examples are the growing support for the Concentrated Solar Panels (CSP) and electric vehicles in the USA (Gillingham & Stock, 2018). However, whether these considerations become realities is uncertain.

The cost-reducing technical change has been explained by progress in research and development, specialisations, and other factors within the production of firms called the 'endogenous factors'. Another group of factors addresses changes outside the production called 'exogenous factors' which can be larger policy support, or wider consumer demands that generate more sales. An illustrations is in PV. The cost-reducing change of PV shortly after its introduction in the 1970s was driven by endogenous factors. An important driver was an exchange of knowledge between experts in the innovating companies, the so-called 'knowledge spillover'. These spillovers emerged because experts changed jobs, vested own firms, and adapted novelties based on publications, which generated a 15% cost-reducing technical change from 1974 to 1999 in Japan (Watanabe et al., 2002). These endogenous changes were largely due to higher effects because more light-sensitive elements are compounded in the PV cells, referred to as 'the law of Moore'; it predicts reduction in the unit costs by half, for every doubling of transistors on the integrated board per period of time as observed in micro-electronics and photovoltaic cells. Note that the exchange of expertise rather than secrecy drove this technological change. However, this was not sufficient to keep the cost-reducing technical change. Japan gained from the first-mover advantage due to the knowledge spillovers, but the domestic market remained small scale because other energy resources are prioritised, particularly nuclear power expanded. This choice permitted the PV producers in other countries to catch up with the Japanese producers because they produced PV panels on large scale for telecommunication, housing, and other private markets; as well as for military uses, and other public markets. After the 1990s, the Japanese producers lost the PV market; inter alia, they were also unsuccessful in the nuclear power because the global market stagnated as the production costs rarely declined but often increased. These exogenous changes generated even faster cost-reducing technical changes due to economies of scale, referred to as 'the law of Wright' which predicts lower unit costs in manufacturing per period of time as the scale increases. Based on that latter period in development of PV, it is assumed that the exogenous factors dominate in the PV technology (Doyne Farmer & Lafond, 2016) but this conclusion can be biased by the focus on the later assessment period in the PV development.

The cost-reducing technical change continues for decades because those cost-reducing actions in the production of energy technologies, called 'learning-by-doing', are followed by adaptations of installation during uses when consumers find ways to reduce the costs and improve performance, labelled as 'learning by using'. While the latter is observed, not many studies on energy are found. For example, a study shows a 20–30% cost reduction in desulphurisation of vent gas in the USA power sector during several years after the installation of equipment (Wiersma, 1989, p. 230–235). A statistical study into desulphurisation and denoxing caused by combustion of fossil fuels in the refinery, chemical, metal and electricity sectors in the Netherlands from 1985 to 2002 has shown the cost reductions by 17%, to 23% a year. Fast cost-reduction during a few years after the installation of those technologies was usually followed by the decreasing rates of the annual cost-reduction (Krozer, 2008). Learning by using can also invoke innovations. For example, uses of the PV panels combined with transformers, batteries and other accessories invoked stand-alone systems which compete with the diesel generators.

In theory, more effective energy production should reduce carbon intensity of economies. However, this is not observed. An engineering assumption is that economies evolve towards a larger use of energy resources with higher hydrogen to carbon ratios because hydrogen determines the energy density in fossil fuels (Herman et al., 1989). For instance, nearly twice more hydrogen to carbon in natural gas compared to coal can deliver a nearly twice higher energy output in production; similar is argued about a higher electron density in nuclear resources than in solar ones for power generation. Based on this argumentation, shifts in global energy consumption from carbon-rich resources to hydrogen-richer ones are assumed, referred to as the 'decarbonisation of economy'. The shifts from peat and biomass to coal in the 1800s; then, coal to oil and gas in the 1900s; as well as from hydropower to electron denser nuclear power are supposed to illustrate the decarbonisation (Grübler & Nakicenovic, 1996). Thus, lower energy density of modern renewable energy is considered an anomaly on the energy market driven by subsidies (Hirth, 2013). However, the trend toward lower carbon to hydrogen ratio is not observed in energy consumption. Measured by global CO_2 emissions during the last century, the emissions increased in total as well as per energy unit. What is observed is the saturation of growth mainly due to the growth of nuclear power and modern renewable energy driven by the expanding electricity consumption.

That difference between the production and consumption of low-carbon energy resources can be explained by the allocation of innovation-rents which means profits from innovations. Novel energy-efficient technologies that save costs generate innovation-rents. If the innovation-rents are allocated into additional energy consumption, the access to energy is increased while the energy efficiency in energy production is countered by larger energy consumption. This is called the 'rebound effect' or 'Jevon effect', named so after the nineteenth century English engineer and economist who observed that coal consumption grew when coal supply became more efficient. Although rebound effects are often perceived as failures from the environmentalist viewpoint as they impede emission reduction, they can increase welfare when valuable qualities are added. For example, the rebound effects of better and cheaper lighting hardly reduced energy consumption for lighting, along with the increased productivity and better education, care, and other consumptive activities that contribute to welfare (Fouquet, 2011).

Based on the global conversion of primary energy resources into consumable energy, it is estimated that global energy efficiency increased from about 18% in the early 1800s to 40% in 2015 (Fattouh et al., 2019). If this estimate is right, which is plausible based on research into power generation, and energy consumption for lighting and other sectors, the effect-increasing technical change of the global economy is about 0.4% after corrections for the rebound effects. Meanwhile; the global real income grew about five times faster which implies substantially larger positive effects of the rebound effects on welfare compared to the effects on energy consumption. Therefore, welfare can grow alongside lower pollution if the innovations-rents cost-savings are allocated into higher energy efficiency and larger renewable energy.

The energy density of resources is rarely the key factor in energy consumption because the main mechanisms of change are effect-increasing with cost-reducing technical changes. They generate welfare growth along with pollution reduction if innovations-rents in energy production are allocated in energy efficiency and renewable energy, though the drivers of these allocations are insufficiently explained by the evolutionary theory.

3.5 Valuable Energy Services

The value added by the energy services is another driver of change because enables to increase profits of producers and generate benefits by consumers. This mechanism of change is labeled as the valorisation of energy services. Studies on energy consumption in England indicate that scarcities of energy resources caused higher prices from the 1300s to 1700s. This trend turned into decreasing prices during the 1800s and 1900s. This turnaround in prices is attributed to the decreasing costs of energy services. While the annual prices of energy services decreased on an average of about ten times faster than the annual resource prices across sectors, the decrease was 1.3 times faster in power, a few times faster in heating and transport, and nearly 650 times faster in lighting (Fouquet, 2011). Cheaper energy services generated higher productivity, which enabled high income. The higher income invoked an even larger energy consumption as a unit additional income triggered more than one unit of energy consumption – the price elasticity of energy demand was low while the income elasticity was high (Fouquet, 2016). That income elasticity enables purchases of costly innovative energy services if they provide benefits in consumption, and as the services disseminate, the unit costs decrease and prices go down driven by competition between services providers. The benefits in consumption refer to functional and ethical qualities which are valuable to consumers and generate novel businesses, called 'spin-off.

Innovative energy services were purchased despite higher prices. For example, huge expenses were made by households in the 1800s England when open fireplaces in houses were replaced by less smoky and energy-efficient furnaces with chimneys, which reduced health hazards and prevented fires (Allen, 2012); similar changes are observed in low-income countries in the 2000s. The status also invoked innovations. For instance, the early automobiles of the mid-1800s were purchased to show wealth as their performance was lower than that of stagecoaches; inter alia, there are no fewer horses by the 2000s, mainly for leisure (Horses 2021). Public security is also an important driver of better energy services. The electric city lights in the late 1800s were costlier and less effective than coal gas lights, but prevented city fires. Societal engagement is relevant, as well. For instance, biodiesel used by farmers for tractors was refined to fuel for the car drivers aiming at the mitigation of climate change in the late 1900s; the drivers advertised it as 'vegetarian driving'. Those cases illustrate that energy services can be costly for consumers when introduced but purchased for various functional and ethical qualities entailing the effect-increasing and cost-reducing technical changes. Costly solar panels are also introduced thanks to the trailblazing purchasers that pursued higher status of a carrying firms or household.

In addition to those consumer benefits, the innovative energy services generated novel businesses. This spin-off was observed in various sectors. Many examples can be mentioned but systematic studies on the spin-off of innovations in energy business are not found. Better conversions of fossil fuels in the early 1800s generated more powerful rolling in steel mills and higher heating for the cement industry; development of chemical energy in the mid-1800s enabled electricity services off-grid in batteries; followed by a wide range of businesses in communication, exploration of radioactive energy in the mid-1900s triggered military services as well as the nuclear medicine. The electricity services on-grid and off-grid fostered fine mechanics and electronic which diversified services into all kinds of data processing, in industries, mobility, and leisure. In the late 1900s, modern renewable energy enabled off-grid operations in telecommunications, exploration of space, and other services. Those novel energy services were costly when they were adopted. Despite of high prices, they were sufficiently attractive in market niches, followed by scaling up in production during diffusion along with decreasing costs. The adoption of beneficial, high-price innovative energy services followed by the cost-reducing technical change is an important driver of income growth. For example, an estimate in the USA is that the prices of durable consumer goods during the 1970s and 1980s decreased by an annual average of 3% to 4% due to better energy services, which is a major factor in the GDP growth (Gordon, 1990).

A challenge is measuring those benefits. An indicator is an income per energy unit, called energy performance. The damages -disbenefits- caused by the energy services can be indicated by income per carbon unit, labelled as carbon performance. They are estimated as the GDP per energy consumption and GDP per CO_2 tonne. Higher energy performance and carbon performance indicate higher benefits and lower damages of energy services. The main limitations of these indicators are that they do not differentiate between the benefits accrued within value chains of energy and in other sectors, neither higher performance due to lower costs or better effects of energy services, nor specify benefits of the particular energy service. The energy performance and carbon performance indicate solely the monetary benefits of energy services after the rebound effects, whereas the non-monetary impacts are excluded; for instance, appreciations in societies or distribution of benefits. Hence, these indicators are far from perfect. Nevertheless, these indicators enable comparison across countries and overtime in order to assess how energy consumption, as well as the energy-intensive industries and applications of fossil fuels influenced income growth.

A popular idea is that the growing industries generated income growth. While industries need energy in particular the energy-dense fossil fuels, they gradually improve energy efficiency. In effect, it is argued, the energy-intensities and carbon-intensities of economies, measured by energy and fossil fuels per GDP increased during early industrialization followed by its decrease. This idea is often presented as a reverse U-shape function of energy-intensity of time (Goldenberg & Reddy, 1990). This idea implies that energy-dense fossil fuels are necessary for economic development. It is partially confirmed by historical studies. A study on the industrialisation of England from 1600s onwards shows that its energy intensity increased during the 1700s, followed by a decrease from the mid-1800s onwards due to more energy-efficient technologies; note that the industrialization in England grew nearly a century earlier than in other countries (Warde, 2007). A similar pattern of increasing energy intensity during 1600s and 1700s followed by a decline during 1800s is shown for Western Europe (Malanima, 2014). However, energy intensities also varied across a few European countries between 1865 and 1965. For example, steady declines were observed in Sweden and Spain, little changes in Italy, and fluctuations in the Netherlands (Gales et al., 2007). The regional assessments confirmed different development patterns. Whilst the energy intensity continuously declined in central and northern Italy over four centuries from the 1600s on, it increased followed by a decrease in England (Malamina, 2016). The energy intensity declined steadily during the last two centuries in the USA as shown in Chap. 1.

Those patterns were presumably caused by various economic structures across countries because solely the energy-intensive basic industries needed energy-dense fossil fuels whilst their contributions to the countries GDP's were usually low compared to food, textile, and other consumer-oriented industries, as well as growing services. Apparently, the reverse U-shape of energy-intensity in economic development is not a law of nature but a result of decisions in favour of basic industries that cause energy-intensity economies. Globally, the energy performance – reciprocal to energy-intensity – steadily increased during the last two centuries whilst the carbon performance decreased. Figure 3.6 shows the index of USD_{2011} per kWh and per tonne CO_2 emissions from the year 1800 to 2000 (1800 = 100). Those observations are based on Maddison data for income, OWD data for energy, and CO_2 emissions which are estimated with the IEA-BP coefficients multiplied by the energy resources with OWD data. Note that global energy production equals to energy consumption.

The global energy performance grew on an average of 0.7% a year throughout the last 200 years. That growth rate of energy performance can partly be explained by a 0.4% growth of the global energy efficiency after correction for the rebound effects, as mentioned in the former section. The remaining 0.3% growth of energy

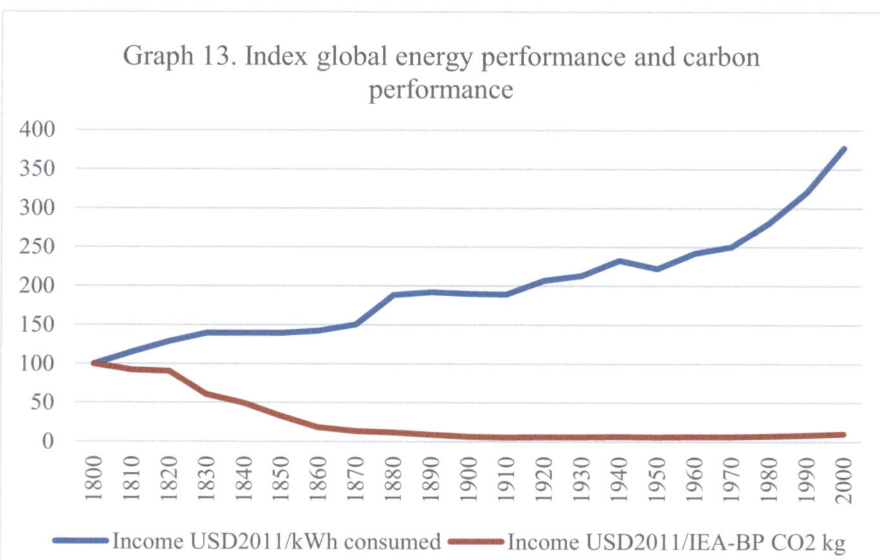

Fig. 3.6 Index global energy performance and carbon performance

performance can be explained by the valorisation of energy services due to the beneficial energy services for individual consumers and spin-off. The growth rates accelerated to about 2% during the economic booms in the late 1800s and late 1900s. The former boom evolved during the declining fuel prices, measured by oil prices, the latter one was during high fuel prices from 1974 to 1985, which indicates that high fuel prices do not necessarily obstruct the income growth but generate far-reaching improvements of the energy performance. The carbon performance decreased during the 1800s as more fossil fuels were consumed. That decrease stopped after the 1920s and turned into a slow increase by the late 1900s when low-carbon technologies based on nuclear power and modern renewable energy grew. While beneficial energy services increased energy performance during economic booms, higher energy efficiency stopped the decline of carbon performance.

The energy performance differed across continents which underpins different patterns in economic development. Figure 3.7 shows this for the period from the year 1930 to 2000 in ten years intervals. Herewith, the GDPs on continents are based on Maddison countries data, and the energy data are derived from the SHIFT database on energy production based on the assumption that the observed production and consumption of energy on the level of continents are equal.

In the early 1900s, the energy performances in North America and Europe were low and steadily increased during the last century. Meanwhile, the energy performances were higher in Russia and Oceania compared to Africa, Latin America and Asia but declined faster. After the 1980s, the energy performances on all continents increased, and converged due to faster growth in income than that of energy consumption. These changes can be explained by a combination of the growing

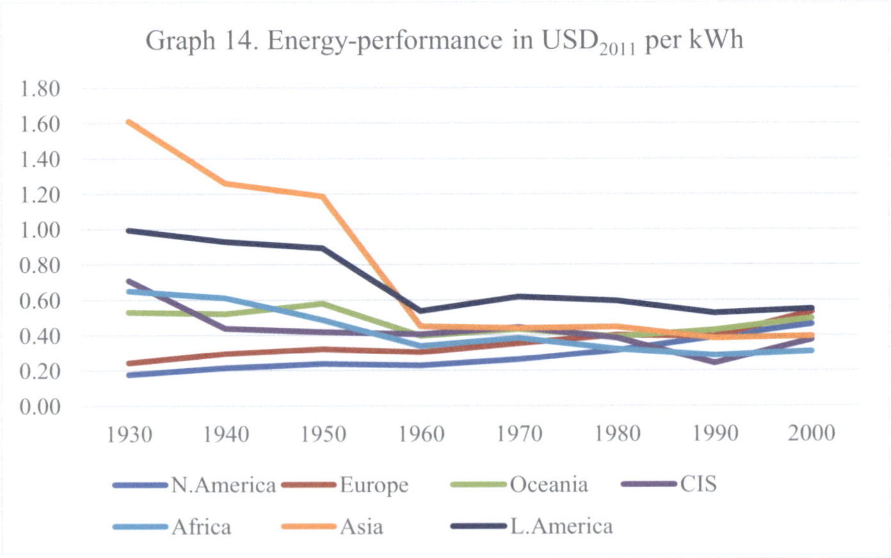

Fig. 3.7 Energy-performance in USD_{2011} per kWh

shares of services in nearly all economies, growing energy-efficiency in production, and the higher value of energy services with modern renewable energy. High fuel prices in the 1980s enabled income growth alongside better energy services across all continents, which underpins that high fuel prices do not preclude better economic performance. Across the continents, the energy performances converged during the late 1900s because lowest energy performance was USD_{2011} 0.31 per kWh in Africa and highest one was USD_{2011} 0.53 per kWh in Latin America compared to a ten times larger spread in energy performance in the early 1900s. Nevertheless, it is observed that the energy performance is generally higher in high-income countries than in low-income ones (Brito & Sausa, 2015). Along with the converging performance, energy consumption per capita also became better accessible across the continents because converged from 300 times higher in the USA than in Latin America in 1930, to 8 times higher in the USA than in Africa in the late 2000s.

However, the carbon performance did not converge in those 70 years. Figure 3.8 shows the carbon performance on the continents in USD_{2011} per tonne CO_2 for the period 1930–2000. The same data as in the assessment of energy performance is used; in this case, the energy resources are multiplied with the IEA-BP factors for the CO_2 emissions.

The carbon performances in North America and Europe hardly improved from 1930 to 2000, which means that the environmental damages continued. The performance of Russia declined during the first half of the 1900s due to a larger scale of energy consumption based on coal and oil, which implies larger damages. This is followed by an improvement mainly due to the growth of gas rather than coal and oil. It declined in all other continents until the 1980s, and particularly fast in Asia and

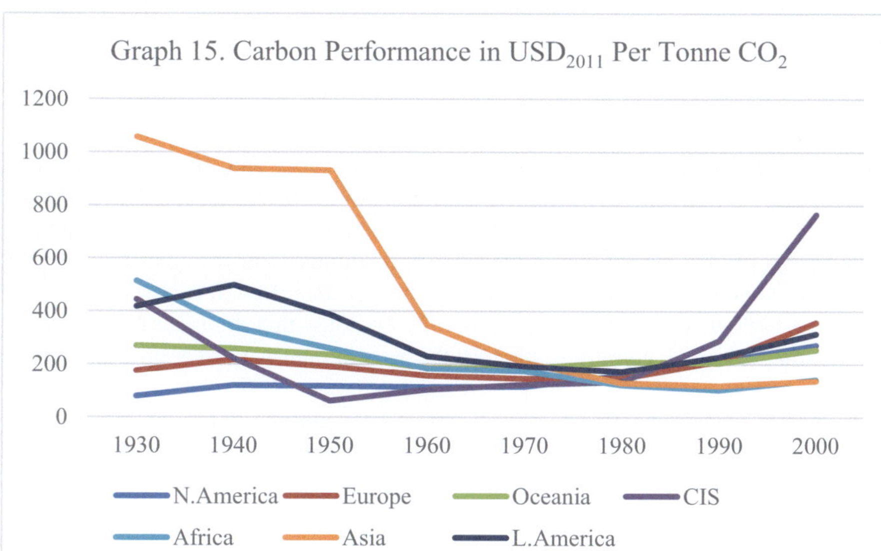

Fig. 3.8 Carbon performance in USD_{2011} per tonne CO_2

Africa caused by the growing consumption of fossil fuels compared to biomass. During high fuel prices in the 1980s, the carbon performances improved on a few continents, not in Asia and Africa. The carbon performances converged across the continents at a lower level until the 1980s, implying greater environmental damages. Thereafter, it diverged as the Asian and African carbon performances only slightly improved. The divergence is mainly caused by the shift in energy resources from biomass to fossil fuels in many mid- and low-income countries. The shift from traditional renewable energy to fossil fuels evolved during the late 1900s but modern renewable energy hardly grew. This pattern of global changes poses a challenge for the coordination of international policies on the mitigation of climate change. Even high of fossil fuels in the 1970s and 1980s hardly improved carbon performances, except in Russia due to gas consumption. Therefore, positive effects of higher fuel prices through global CO_2 taxes on income and environmental qualities can be expected only if such tax is high, higher than the CO_2 equivalent of high oil prices.

Concern remains about the impediments for higher energy and carbon performance despite high fuel prices. An observation is that large producers of fossil fuels attract the best capabilities through high salaries, policy support, and other means because they generate large-scale activities with high profits. Therefore, it is argued, the producers of fossil fuels detract from the valuable capabilities in other sectors, labelled as the 'curse of natural resources' (Sachs & Warner, 2001). This observation is tested based on the growth of energy performances of 61 countries and the world. The growth rates are estimated using Maddison data for GDP and SHIFT data for energy production in ten-year intervals based on the averages of the preceding ten years, whereas a few extreme data from the early decades of the 1900s are excluded because presumably riddled with imperfections. The countries' data are divided into the 1st and 2nd half of the 1900s because the violent 1st and 2nd World Wars, and the liberation wars from colonial powers in Asia in the first half of that century are followed by the less war-like second half and faster income growth. Appendix 2 shows indicators for 61 countries and the world. The indicators cover: per capita income and energy production in 2000, energy performance in 1900 and if data are unavailable the performance in the year closest to 1900, and in 2000, the annual average growth of energy performance during 1900–1950, as well as 1960–2000 and 1991–2000. Countries that are net exporters of energy resources are considered to be large energy producers; net export means exports minus domestic consumption plus import. Fourteen countries are estimated to be net exporters. It is assessed if they generated high energy performance. It should be noted that the income and energy production grew in all countries, whereas the growth rates vary over time and across these countries.

In the 1st half of the 1900s, only 8 out of 61 countries improved their energy performance, whereas 29 countries improved their performance during the 2nd half when incomes grew fast. Those countries whose performance improved in the first half performed less in the 2nd half. As a result, faster income growth improved energy performances in most countries. Correlations are estimated between the countries' growth of income and energy production in the years 1920 to 2000; earlier data on many countries' energy production are deficient. High correlations ($R^2 \geq 0.8$) are found only for 10 out of 59 countries, whereas data on the two

Table 3.5 Patterns in energy performance; bold are net exporters of energy resources (production higher than domestic consumption) based on the SHIFT data

Criteria	Increase	Decrease
Above 1 USD_{2011}/kWh	Austria, Belgium, Chile, **Germany**, France, Ghana, Hungary, Ireland, Italy, Japan, Korea, Morocco, Portugal, Spain, Switzerland, Taiwan, Turkey, Yugoslavia	Cuba, Finland, Greece, Jamaica, Jordan, Lebanon, Philippines, Sri Lanka, Tunisia
Below 1 USD_{2011}/kWh	Bulgaria, Czech & Slovak Republics together, **Iraq,** Netherlands, Peru, Poland, Romania, **Saudi Arabia**, UK, USA, **Venezuela**, (World)	Argentina, **Australia, Bolivia**, Brazil, **Canada**, China, Colombia, Denmark, **Egypt**, India, **Indonesia, Iran, Mexico**, Myanmar, New Zealand, **Norway**, South Africa, **Soviet Union,** Sweden, **Syria,** Thailand, Uruguay, Vietnam

countries are insufficient. High ones are found in the European high-income countries. However, high income is not the sole factor regarding low correlations in Australia, Canada, Japan, the USA, and other high-income countries. Given the moderate correlations, all countries are divided into quadrants. This is based on high energy performers above one USD_{2011} per kWh in the year 2000 and low performance below that, as well as the increasing or decreasing energy performance during high-income growth from 1960 to 2000. Table 3.5 shows the results. The countries shown in bold are large energy producers based on net exports of energy.

Nearly half of all countries – 29 out of 61 – improved their energy performance during the last four decades of the 2000s, which underpins the fact that the decoupling of income from energy is widespread. However, only 4 improving performers were large energy producers which suggests that a larger scale of energy production rarely contributes to income growth or energy efficiency. Nearly half of all countries – 27 out of 61 – are high energy performers, whilst only Germany among them was a net energy exporter in the period 1960 to 2000. Among 34 low performers, 13 are net energy exporters. This means that large energy exporters were generally low energy performers. Moreover, energy performance of the low performers declined throughout the 70 years implying that the efficiencies declined as the energy exports grew. These observations support the hypothesis that large production of natural resources impedes innovations in the energy resources, thereby generates low energy performance along with low carbon performance. This result indicates that changes in the energy intensity are not quasi-autonomous economic developments caused by industries. The changes are largely driven by the interests in energy production, which may generate high profits within the energy sector but often impede the valorisation of energy services, thereby often obstruct the income growth.

Energy services improve qualities in energy consumption and create spin-offs. This improves energy and carbon performance. Increasing energy performance is observed in all continents, whereas the carbon performances diverge. High energy performance is mainly driven by valuable energy services whilst large, vested interests in energy production obstruct that performance.

3.6 Conclusions

Mechanisms that have changed the consumption of energy resources during the last centuries are assessed. It is observed that global energy consumption was mainly driven by a larger population due to declining mortality when growing income per capita in the late 1800s enabled better and longer lives of many people, and that the growth of energy consumption is linked to the energy-intensive activities which are not the income-generating ones per se. Whilst the incomes grew faster than energy consumption in North America, Europe, and Oceania, they fluctuated in Russia, and grew slower in Africa, Asia, and Latin America. Faster growth of income compared to energy consumption was mainly driven by the growth of services whose share increased throughout 200 years on all continents, whilst that increase evolved faster in North America, Europe, and Oceania than on other continents. These shifts to services generated income growth along with a slower growth of energy consumption and CO_2 emission.

Within energy consumption, the shift from biomass to coal during the 1700s can be explained by higher prices of scarce wood, given growing demands for energy. However, the expansion of oil in the late-1800s, hydropower in the early-1900s, natural gas, and nuclear power in the mid-1900s cannot be explained by the scarcities, and prices as the production of fossil fuels grew while the real prices declined slowly until the late 1980s. High fuel prices in the late 1900s and early 2000s were driven by the international policy agreements and growing energy consumption. An unintentional effect of high prices was the emergence of modern renewable energy. High fuels prices triggered biofuels, wind, and solar power which continued to grow during low prices. This suggests that high fuel prices can invoke shifts towards new energy resources while their growth depends on other drivers as well.

High fuel density is a minor factor in the changes compared to the conversion technologies for higher energy efficiency and lower costs in life cycles of energy services, it is the effect-increasing and cost-reducing technical change. During the 1900s, a few percent effect-increasing and cost-reducing technical changes in the global energy production enabled higher incomes, which are partly allocated in larger energy consumption and partly in labour for higher-value services. In effect, the energy efficiency in consumption increased only by a 0.4% annual average despite nearly tenfold cost-reducing technical change.

Innovative energy services add beneficial qualities in energy consumption and generate novel, higher-value activities across sectors called spin-off. Value addition by the energy services is indicated by higher income per energy produced, it is energy performance. The global energy performance increased by 0.7% annual average. Large differences in the energy performances across continents are observed in the past, but convergence evolved towards higher energy performances. Meanwhile, the carbon performances, which indicate the damages of energy consumption, hardly improved during the 1900s and diverged across continents in the late 1900s because many mid and low-income countries consumed more fossil fuels.

Experiences during the 1900s indicate that improvements of higher energy and carbon performances are obstructed by large interests in fossil fuels which do not contribute, but often even impede the countries' income growth.

Global economies shifted from traditional renewable energy resources to fossil fuels during the last two centuries and started to re-shift towards modern renewable energy during the late-1900s. The main mechanisms of change in that re-shift are the growth of labour-intensive services compared to material-intensive agriculture and industries, effect-increasing and cost-reducing technical change in energy, and an increasing energy performance due to higher-value energy services. Meanwhile, high prices of energy resources did not impede global income growth but invoked energy performance. Policies can foster these mechanisms by shifting taxes from labour into energy resources, and support value addition in energy services.

Chapter 4
Inventions in Renewable Energy

Can innovations in renewable energy be forecasted and steered? Historical highlights in basic energy technologies along with statistical data are employed to assess lead-times of energy innovation processes from the past, present chances of innovation in renewable energy, and financing possibilities. The highlights of 29 technologies show that lead-times of innovation processes – from inventions to innovations and diffusion – usually cover six to ten decades. Lead-times are less than six decades when driven by the military interests and more than ten decades when the applications cause harm to health and the environment. Data on the R&D and patents during the late 1900s and early 2000s across countries indicate larger business interests in renewable energy than in other energy sectors, in particular when combined with information and communication technologies into the distributed energy systems. However, predicting profitable innovation based on R&D projects are shots in the dark unless policies address particular challenges. The possibilities of profitable investments in renewable energy increase due to the policy demands and consumers' participation in innovations.

4.1 Introduction

It is discussed how innovations in renewable energy are pursued and can be forecasted and enhanced. In the conventional line of reasoning, innovations are considered successful when research & development (R&D) delivers novel concepts (inventions) which enable the start-up of firms (innovators) that generate net benefits during the dissemination of inventions (diffusion) which enable to operate in the competition and contribute to welfare. These thoughts about innovation processes are embedded in institutions to garner public support for R&D and protect inventions by Intellectual Property Rights (IPR), called 'innovation system'. That public support of R&D encompasses public expenditures on the firms' R&D as well as subsidies, credits, guarantees, and other instruments aiming to foster firms' start-

Y. Krozer, *Economics of Renewable Energy*,
https://doi.org/10.1007/978-3-030-90804-1_4

up toward profitable maturation. Whilst the public results of R&D are measured by publications, private ones are estimated by the IPR because assumed decisive for successful firms. Within IPR, patents are focused on technological concepts, whilst trademarks, models, designs, and copyrights address mainly commerce, informatics, and arts & crafts, respectively. In high-income countries, mainly patents are used. For example, patents covered 59% of all IPR in the USA and 68% in Japan during the period 2008–2017 but only 3% in Bangladesh and 1% Ethiopia where trademarks are more popular. Therefore, R&D that delivers patents is assumed to indicate business interests in technologies whereas, the success of innovations is measured by the birth of firms and survival of start-ups after several years of competition. Questions are if the innovation systems in energy are effective, particularly if private investors and policies can assess what R&D should be financed and subsidised in order to generate innovative start-ups and whether the patents in energy are important for the survival of the innovative start-ups in competition. Answers to these questions are relevant because ultimately the survivors generate benefits that can cover the costs of R&D and justify the IPR's.

The justification for the innovations system is that most R&D is based on private funds which must be recovered by profits while research is largely funded from public sources. The statistics of the Organisation for Economic Co-operation and Development (OECD) show that 72–78% of all R&D during the 2000s in the OECD countries has been covered by enterprises. Such data on energy R&D is scarce though 60–70% of it in Korea and Russia are paid by enterprises. By far most energy R&D is done by the consuming business, mainly industries, as only 0.1% and 0.4% of the energy R&D are paid by energy producers in the USA and the Korean Republic (Korea). That expenditure is perceived effective if many patents are approved by the patent office. The patent office in turn vests ownership on technology for a period of time if the inventors pay a fee. This way, the owners of patents possess the authority for the use of the patented technology and licensing of other users, at a price. This enables the protection of the patent-holders from unintended copies that are deemed illegal, and avoid disputes about the ownership. Although patents are often used by scholars as indicators of business interests, this ownership is rather a signal of high status in the scientific community. While many inventors pay for holding their patents they do not start firms and many successful firms operate without patents though patents are popular indicators of the innovative activities (Grilliches, 1996).

This idea about an innovation system is widely accepted in economics; what differs is the assumption about causalities and the role of public funding. Mainstream, neoclassical economics assumes that high market prices signal scarcities, thereby trigger R&D for cheaper technologies, and if successful, the patent-holders vest monopoly during the validity period of patents because competitors may not duplicate these concepts at the risk of fines. The advantage of the monopolist is that the firm can demand high prices for applications of its novelty despite low production costs if the novelty deemed attractive to customers (Stoneman, 1983).

From the evolutionary perspective, inventions are developed based on cumulative knowledge from various sources, followed by a selection of the novelty by producers whose decisions are bounded by political, cultural and other conditions, which caused imperfections (Colinsk, 1996). Imperfections in decisions about energy technologies are observed for example, when the damages caused by polluting energy consumption are deficiently priced and improvement is neglected. Therefore, the idea that policies should generate and steer technologies through public funds for the realization of public objectives is widely embraced in the evolutionary analyses of policy-making on energy innovations (Arentsen et al., 2001).

The behavioural argumentation is the R&D is based on knowledge that is loosely related to a particular innovation process because innovation is usually driven by entrepreneurial tinkering; it is experimentation aiming to resolve a problem with short-term actions. It is observed that most inventions, whether patented or not, are generated by the experimentation of entrepreneurs who may have failed in business, thereby forgotten in history, but whose novelties are stepping stones for the inventions of others. This is observed in inventing utensils in daily life (Petrovsky, 1994) as well as for complex systems such as airplanes (Petrovsky, 1996). The evolutionary idea about the innovation system is disputes, in particular the role IPR is debated. While patents protect the inventors, they also impede innovations because vest monopolies unless the duration and scope of patents are restricted. Restrictions on patents are advocated by the proponents of free-market (The Economist, 2015) and development economics (Al Musa & Khidzir, 2018).

In the behavioural argumentation, the entrepreneurial capabilities in assessing business opportunities, experimentation with technologies and generating resources, are assumed the drivers of innovation. These capabilities are generated through knowledge spillovers, stakeholder interactions, social engagement, and other interactions rather than formalised R&D (Shane, 2000). Based on observation of the entrepreneurial behaviour, it is argued that successes depend on those capabilities rather than on policies, finances, and assessments of alternatives (Sarasvathy, 2001). Instead of the assessments aiming to reduce risks of innovations, which fail because uncertainties about future events cannot be avoided, it is advocated to foster the entrepreneurial capabilities in the converting of observed deficiencies in markets into solutions that are demanded by customers and societies on a whole (Taleb, 2012). Those notions encouraged ideas about enhancing the knowledge spillovers; for example, hubs for startups, social networks, meeting points and other exchanges of know-how.

Herewith, possibilities for innovations in renewable energy are assessed mainly using the mainstream argumentation. The lead times in the processes of energy innovations are assessed based on time lags between inventions, innovations, and diffusion. Business interests are estimated using statistical R&D expenditures and patents in renewable energy, when compared with overall energy, and all sectors. The chances of successful innovations in all businesses are compared to the chances in renewable energy and the financing of innovations, in general and in renewable energy are addressed. Finally, conclusions are drafted.

4.2 Lead-Time in Energy Technologies

The time that elapses between the proof of a technological concept (invention), the start of a firm based on that invention (innovation) and the dissemination of the invention driven by large-scale sales (diffusion) is the lead-time of an innovation process. The lead times of all basic energy technologies in the last two centuries are assessed; a basic energy technology means technology that added an entirely new energy product. However, production technologies are largely neglected. The lead times are assessed with the literature reviews (Podobnik, 2006; Quaschning, 2008; Smil, 2019), and references to specific technologies. The inventions, innovations, and diffusion are highlighted and ordered in accordance with the subsequent introduction of energy resources over time, which spans biomass, coal, electro-chemical, oil, gas, hydro, nuclear, geothermal, wind and solar. There are several limitations in the assessment. First of all, only successes are listed. Second, these timelines cover only basic energy technologies for the market applications, it means excluding non-market uses of biomass, energy for human and animal labour in foods. Third, the selection of technologies addresses resources, and products. This implies that the exploration, exploitation and conversions are linked to resources, and products; for example, deep-water drilling and catalysts to oil production. Fourth, non-conventional energy resources and immature technologies are not covered. Finally, this assessment is focused on the USA and Europe, which have been the front-runners in shifts from traditional renewable energy to fossil fuels, and modern renewable energy. Innovators in other countries could have been global leaders in particular technologies; for example, the Brazilian processors of sugarcane for bioethanol and Japanese electronic firms in PV.

Based on this selection, 29 basic energy technologies are identified which is a large number compared to, for example, 19 basic technologies as pinpointed by a Russian assessment (Mitrova, 2019) even less in other reviews. For the convenience of the readers who prefer to skip the brief descriptions of basic energy technologies, the lead times of technologies are shown in Table 4.1. In this table, the timeline is presented horizontally, and technologies vertically. The inventions are shown in italics, innovations in normal letters, and diffusion in bold.

4.2.1 Bioenergy

Bioenergy is energy based on biomass. It is burned or processed into bio-alcohols, bio-oils, and biogases. Burning biomass covered 98% of the global energy consumption in 1800 while 11% in the year 2000, equal to 12.5 PWh; which is small compared to 3000 PWh biomass potential based on 550 billion tons of timber, straw, fat, and other biomass production in a year. The growth of bioenergy consumption is saturated and declines per capita (Fernandes et al., 2007). Four main resources are addressed: wood, bio-alcohols, bio-oils and biogas.

Wood was the main energy resource for many centuries. Mostly, it was directly burned or pressed, or heated in low-oxygen conditions (pyrolysis) for the energy denser briquettes and charcoal, respectively. These are important energy resources for people with low incomes in low-income countries, but poorly statistically reported (FAO, 2016). Fuelwood is gradually outcompeted by coal during the 1800s, despite innovation in the distribution of wood; for example, the introduction of wood chippers in the USA in the late 1800s for higher density of fuelwood in

Table 4.1 Timeline of basic energy technologies, ones that generated new resource; inventions are shown italic, innovations – normal, dissemination – bold; Elect-electrical resources, AC-Alternating current, DC- Direct current, based on own estimates

	Bio	Coal	Elect	Oil	Gas	Hydro	Nuclear	Geother.	Wind	Solar
(before1800)	**Burning oils**	**Coke gas** (Steam engine)	*Batteries (direct current)*	*Tar sand (seeps)*	*Nat.gas (seeps)*	**Small scale**	*(Mine hazards)*	**Heating**	**Mech. anical**	**Heat**
1810	*Distillat. Alcohol*									
1820		**Steam engines**			Gas delivery					
1830			*Current AC*							
1840			*Fuel cell*	*Oil well drilling*						
1850			Batteries (DC)	*Refinery kerosene*						
1860			*Batt. Recharge*	Oil wells drilling	*Fracking*					
1870	Bio-alcohol	*Boilers*	Lamp bulb	Gasoline 4 stroke						
1880			Power plant AC	*Tar sand*	*Heating Bunsen*	*Water turbine*			*Horiz. Rotor*	
1890	*Biogas*		**Batteries recharge**	Diesel ignition						*Solar heater*
1900		Boilers firm		**Oil well drilling**	Gas heating	Power plant		*Geoterm electric.*	Horiz. Rotor	
1910	**Bio-alcohol**			**Diesel movers**	*Thermal reform.*	Various turbines				*CSP*
1920		*Fluidized bed comb*	**Power networks**	**Gasoline cars**	Pipe lines					Solar boilers
1930	Bio-diesel	**Boilers**	Fuel cell	Tar sand	Thermal reform.	**Large scale**			*Vertic. Rotor*	

(continued)

Table 4.1 (continued)

	Bio	Coal	Elect	Oil	Gas	Hydro	Nuclear	Geother.	Wind	Solar
1940			**Batteries recharge**	**Aviation kerosene**	Fracking		*Fission contin.*	*Heat transfer*		
1950					**Pipe lines**		Power plant			*PV* continue
1960		Fluidized bed comb	**Fuel cell**		**Gas heating**		**Expan sion**	Geoterm electric		CSP
1970			*Laser*		**Therm. Refom.**				Horiz. Rotor	**Solar heater**
1980	**Bio-diesel**	**Fluidized bed com**		**Satura-tion oil**			**Satur ation**	Heat pumps	Vertic. Rotor	**CSP**
1990	**Biogas**		**Lasers**		**Frack ing**				**Horiz. Rotor**	**PV**
2000				**Tar sands**						

transport and higher energy efficiency of combustion. Despite scarcer wood resources, it is still widely used in rural areas for heating and for co-firing in power plants when wood residues are abundant or wood combustion is promoted and subsidised with regard to policies on climate change. As wood became scarcer during the 1700s, more agri-residues for energy are used, thanks to higher agricultural productivity due to better ploughing and harvesting equipment; for instance, wheat in England doubled to 1.7 tonne per hectare in the 1800s starting from 1700, and 2.1 tonne per hectare in the 1900s (Apostolides et al., 2008). Growing agri-residues triggered several conversions for the energy products.

Though the fermentation of agri-residues and distillation of alcohol were known for many centuries the continuous distillation was invented in France in 1813 by Jean Baptiste Cellier, and triggered large-scale industrial production 50 years later, when the invention of the internal-ignition engine emerged in automobiles. The automobiles initially ran on the ethanol-turpentine blend in the USA. When the production of mineral oil expanded in the late 1800s, the bio-alcohols were heavily taxed and bio-alcohols were increasingly replaced by gasoline from mineral oil in the early 1900. In other countries, that replacement was slower and bio-alcohols were used much longer for cars; they are still much used for cars in Brazil, Sweden, and other countries where agri-residues are abundant and exempted from taxes whilst petrol is taxed.

Oils and fats were used for lighting over many centuries; lighting was an important driver for hunting whales. The use of bio-oils increased when internal diesel engines were introduced for fast-movers in the late 1800s, but after 1930, bio-oils were outcompeted by diesel fuel from mineral oil. In many countries, bio-oils re-appeared and grew during the years of high oil prices in the late 1900s, such as rapeseed-methyl-ester (RME) in Germany and refined palm oil in Malaysia when their uses in cars were exempted from taxes and subsidised in order to encourage local agriculture and processing industries.

Biogas is the methane-rich gas generated through anaerobic digestion, which means oxygen-free biodegradation. Indian farmers produced biogas for centuries through the digestion of dung (manure) mixed with the agri-residues for a higher biomass density (Pandyaswargo et al., 2019). Industrial anaerobic digestion was first applied in India, in the 1860s. It reached Europe in the 1890s and remained small scale for nearly one century. In the late 1990s, the production of biogas from agri-residues increased on farms when manure is mixed with agri-residues; for instance, in Denmark, Netherlands and the USA. Biogas is also produced with heat and pressure in the Fischer-Tropsch process patented in 1930. This process is scaled up only in a few countries, e.g., in Malaysia, because it is costly when compared to natural gas. On small scale, biogas is also purified from water, sulfur, and other compounds for a higher density of methane, thereby higher energy-density.

The traditional bioenergy production is called '1st generation bioenergy'; the industrial production from organic residues of agriculture, industries, and households in biorefineries is labelled as '2nd generation' and the biorefining of algae, yeasts, and other micro-organisms for energy and chemicals, which emerged in the early 2000s, is branded as the '3rd generation', yet remains small scale.

4.2.2 Coal Conversions

Coal is a chemical mix of carbon, hydrogen, oxygen, and traces of other compounds in varying densities. The density of hydrocarbons defines the coal quality because it is decisive of its heating value. The main qualities are peat below 3000 cal/kg, lignite 3000–4000 cal/kg, hard coal (sub-)bituminous 4000–8000 cal/kg, and anthracite above 8000 cal/kg, and an even higher density of carbon is graphite which is hardly combustible. The share of coal in global energy consumption increased, from 20% in 1850 to 55% in 1920, when energy consumption grew tenfold. Thereafter, its share declined to 23% alongside a three times larger coal consumption toward the year 2000 when 27 PWh was consumed. The main advantages of coal for energy consumption are large reserves dispersed across many locations and a twice higher calorific value than that of bioenergy.

When the availability of wood declined in the early 1700s, the demands for coal increased along with higher prices of coal. Conversion technologies were also introduced; among them heating coal in oxygen-free conditions for the production of cleaner and energy-denser cokes and coal gas. Cokes are applied in large foundries for the production of pig iron that is used for the construction of machines, whereas gas is used for lighting in cities. The consumption growth of coal is driven by the industrialisation. A larger demand for coal in metal, building materials and other basic industries triggered deeper mining, which needed more powerful exploitation technologies and de-watering methods than the use of horses and people. Their labour was gradually replaced by low-pressure steam engines fuelled by coal, invented in England by Thomas Newcome in 1712. The use of steam machines is accelerated in the late 1700s when the more energy-efficient high-pressure steam engine was invented – by James Watt in 1776 – and widely disseminated due to adaptations tuned to various mining conditions across England and France (Nuvolari, 2010). Coal production in mines expanded when dynamite was patented in the mid-1800s, by Alfred Nobel in Sweden, and disseminated for the mining and warfares. Global diffusion of the steam engines evolved along with the theft of patents and copies of designs, while England remained the leading producer of coal and steam engines until the early 1900s when the USA took over global leadership.

In coal mining, pig iron production, and other heavy industries large-scale, coal-fuelled steam engines were used with a rotation speed of a few times per minute. They were tuned to these large-scale energy-intensive industries. Small-scale steam engines which rotated a few thousand times per minute were patented, in 1884 by Charles Parson, that's 100 years later than the large-scale ones. They were widely used on ships, trains, and other fast movers until the 1930s when they lost the competition to internal-ignition engines based on the combustion of mineral oils. The combustion of coal for generating steam kept growing after the patenting of steam boilers by George Babcock and Steve Wilcock in 1867 in the USA; they needed about 40 years of adaptations in the construction of boilers, and a merger with the Stirling firm, before they started their boiler business. Their boilers, made of brick and tile, could generate up to 25,000 m^3 steam per hour. However, these brick wall boilers lost the competition to tube wall boilers for combustion of the coal that performed better. Tube wall boilers were disseminated along with adaptations for various uses and scales, after the 1920s. By the year 2000, about 80% of electricity production was based on the combustion of coal for steam generation in the tube wall boilers.

Fluidised Bed Combustion (FBC) of coal is invented by Winkler in 1920 in the USA. It increased the energy efficiency and reduced noxious emission of the steam boilers. After this invention, it took several decades before the business start-up. The first FBC was commissioned only 45 years later, in 1965, and this technology was disseminated even later during high fuel prices and stricter environmental regulations in the 1980s and onwards (Koornneef et al., 2007). The gasification of coal into synthetic gas can reduce costs of transport and handling, but is rarely applied because it is costly.

4.2.3 Electric Resources

The electric resources are secondary resources based on the conversions of primary chemical, electric and magnetic resources into electric power used as an energy resource for various conversions. The electric power is distributed as direct current (DC) and alternating current (AC). The direct current is usually stored as the chemical energy in batteries and fuel cells which are used stand-alone, called 'off-grid' some direct current is stored mechanically in flywheels. The alternating current is usually distributed through the electricity networks for consumption, called 'on-grid'. As electricity cannot be stored without additional technologies, this production must be permanently balanced with consumption in the networks. The main advantage of electric power is that it enables nearly all energy services along with larger flexibility by scale, power, frequency, and other qualities in energy consumption. By the year 2000, electricity resources covered about 14% of all global energy consumption, whilst the remaining 86% was mainly heating by combustion of fuels.

Although direct current in animals was known and used for many centuries, the technology of DC has been first demonstrated in 1745 by the Leiden jar with metal and solution. This was then adapted by Benjamin Franklin in the USA a few decades later. Based on those, a battery was invented by Luigi Galvani and Alessandro Volta in 1800, in Italy. They constructed a pile of paired copper and zinc plates, which were electrodes, with cloth and cardboard in-between soaked in brine to act as the electrolyte. While the process of electricity generation was not yet well understood, those experiments paved the way for the (re-)charging, storage, and use of direct current. In batteries, electrons move from the cathode electrode to the anode one through the electrolyte during use, which is reversed with the inputs of external electricity for recharging. Several types of batteries were introduced during its use in the 1800s, with various metals for electrodes and electrolyte solutions. For example, the rechargeable lead-acid batteries were invented in 1859 and were used in cars starting in the late 1800s, whereas the lithium-ion batteries were invented in 1912 and became popular for power storage in electronics only late 1900s (Krepelkova,

2017). During the 1900s, the batteries gained power, were diversified, and disseminated into the prototypes of the electric car by Porsche in 1890, electric movers from 1912 onwards, equipment for high power output in the 1980s, and other services (Zito, 2010). However, batteries deliver power only over a short distance because of electric resistance (Voltage, 2021). Other limitations for its uses are a short period of storage and limited number of recharges.

The alternating current (AC) is produced when copper wires or other conductive wires rotate in the magnetic field. Such an appliance, called a 'dynamo', generates electricity in a process called 'induction'. Dynamos deliver waves of electrons that oscillate in opposite directions; the oscillation frequencies vary across countries between 50 and 60 kHz. This electricity induction was demonstrated in the 1700s by several scholars. These demonstrations enabled the construction of the dynamo in 1831 by Michael Faraday. Electric light was his first service. It was demonstrated outdoors in France and England in the 1840s, which was a few decades before the invention of indoor light. Light tubes, and later bulbs with various filaments were used. Thomas Edison was credited for the patented invention of the light bulb in 1879; he also patented electric machines, a light distribution system, and started with electricity distribution from the thermal plant in 1882. While he was focused on the DC instrumental over short distances, his former employer, Nicolas Tesla, introduced and patented lighting based on AC in 1891. This invention invoked the construction of distribution networks for electricity. In such networks, the transformers at the power plants elevated voltage to 1000s of Volts (kiloVolts) followed by distribution with several smaller transformer units and ultimately the consumption of 110 Volts or 220 Volts in sockets (Carlson, 2019).

Electric power is also used to generate energy resources, particularly hydrogen. Hydrogen (H_2) is an energy dense resource that can be produced with electrolysis, which is a process that can split water into oxygen at the anode and hydrogen concentrated at the cathode. Therefore, electrolysers are used in a device called a 'fuel cell'. Given that the electrolysis process is energy-intensive while hydrogen is an energy-dense, explosive energy resource; the innovation process was slow. First, the fuel cell was constructed and presented to the public, in England in 1838 by Sir William Groove. It took 90 years before Rudolf Erren applied hydrogen in engines of trucks, buses, and submarines in the 1920s in Germany. However, it could not compete with the internal-ignition engines based on fuel combustion. Large-scale fuel cells were first applied in space shuttles in the 1960s which were funded by public, military expenditures. This application was followed by the development of various fuel cells largely based on public funding; fuels cells can be classified by the speed of hydrogen production in the fuel cells that varies from a few second to minutes (Ortiz-Rivera et al., 2007). Policies in the USA, EU, Russia, China and other countries supported fuel cells, and large expenditures were put into the development of fuel cells, but the commercial successes remained far behind the expectation because of high costs and low energy efficiency. Meanwhile, fuel cells in cars could not compete with the DC batteries though large-scale electrolysis is expected to become more energy-efficient and cheaper in the future (Staffel et al., 2018).

Contrary to the slow progress in the fuel cells, the invention of the laser by Theodor Maiman in the 1960s grew fast. This invention paved the way for combinations of light and electromagnetic radiation. Its key feature is to be able to focus the light waves into a bundle of irradiance that enables lighting, cutting, scanning, welding and other works. Within 30 years after that invention, various types of lasers were produced and massively used for fine mechanics, gaming, medical equipment, videos, and other activities (Hecht, 2010).

4.2.4 *Mineral Oil Conversions*

Mineral oil is composed of hydrocarbons with traces of other compounds. Its share in global energy consumption increased from 1% in the year 1890 to 38% in 2000 when 3700 million tonnes of oil equivalent (t.o.e.), about 15 PWh, was consumed. The major advantages, when compared to bioenergy and coal, are higher energy density and convenient uses.

Extraction of mineral oil was invented and patented in the 1700s in England. However, the oil drillings and production emerged a hundred years later in Azerbaijan, Poland, and Romania in the 1840s. They emerged largely based on the use of local technologies with British capital. Oil production in the USA emerged two decades after these trailblazers and grew faster. The first uses of mineral oil were for lighting; which competed with the whale oil; whales were hunted on the seas, particularly for the sperm oil. Inter alia, the introduction of mineral oil rendered lower prices of the whale oil, which accelerated the hunts due to larger ships and heavier hunting equipment for cheaper hunting, which decimated sperm whales; it is illustrative for market failures in the common goods (Kaiser, 2020). Fires and explosions during drillings, in production and use impeded fast expansion of mineral oil. Oil consumption grew after the patenting of low explosive kerosene by Ignacy Lukasiewicz in 1856 in Poland (Krzywiec, 2018). Thereafter, oil production boomed; for example, Romania that was first to introduce oil statistics in 1857 produced 275 tonnes that year, and more than one million tonnes of oil in 1906. Many businesses in the USA, Russia, Iran, and other countries emerged for the oil exploration and mining, called 'upstream production', followed a few decades later by the thermal refining into oil fractions, distribution, and sales for fuels called 'downstream production' (Craig et al., 2020).

The consumption of oil expanded when the four-stroke internal combustion engine was patented by a few inventors in the 1860s, followed by the construction of the two-stroke engine by Nikolaus Otto in 1877, in Germany. This construction triggered the production of motorcycles in 1885, based on the business co-operation between Gottlieb Daimler and Wilhelm Maybach. Within a few years, it was followed by the production of cars with Karl Benz who also patented two-stroke engines and four-stroke engines. Their production capacity in an artisanal manner was surpassed by the serial production of Ford in the USA, from 1908 onwards. Meanwhile, in 1893, Rudolf Diesel patented a more efficient and durable engine with direct ignition. A few years later, he started doing business in diesel engines. His production delivered engines for trucks, ships, and other prime movers; his business expanded after the English marine switched from the steam engines based on coal to diesel engines in the 1910s. Kerosine was mainly used for lighting until the early 1920s when the demand for kerosine expanded in aviation that took off in the 1920s during the 1st World War.

As the consumption of gasoline and diesel cars increased, more gasoline and diesel were consumed which triggered changes in the refining of mineral oil. More gasoline and diesel fractions in mineral oil are refined due to catalytic refining, which gradually replaced thermal distillation from the 1920s onwards. Catalytic refining used catalysts for oil cracking which generated more than a dozen inventions in 20 years of the Interbellum (1920–1940); these inventions followed quickly each other. Thereafter, subsequent catalysts are introduced in the oil market within 8.5–13.5 years; which means that the time-to-market from invention to introduction of the catalyst in the market took about 20 years, whereas the diffusion took longer and not all innovations grew after the introduction (Enos, 1962).

The production and refining costs of mineral oil declined during the 1900s and the scale of production expanded as oil and gas exploitation in deep wells, and in deep water emerged; furthermore, oil and gas exploration with seismic imaging of earth layers is introduced. When the oil price increased from 1974 to 1985 the production of mineral oil from tar sands and oil shale expanded. Although the patents for these technologies were obtained in 1926, industrial production took off only in the late-1970s. For the winning of oil, chemicals were injected into the underground wells under pressure. The 50 years of time-to-market for these technologies was followed by slow dissemination during the subsequent decades until the international oil prices increased again after 2005. Long-time lags are experienced because these technologies are energy-intensive and polluting, which increases costs of operation and triggers societal resistance to the drilling for oil in tar sands and oil shale.

4.2.5 Natural Gas Conversions

Natural gas, which is methane with impurities, water, carbon dioxide, nitrogen, sulfur and other is produced from deep wells and sometimes captured as the associated gas of coal and oil mining. The share of natural gas in global energy consumption increased from 1% in 1950 to 21% in 2000 that is equivalent of 24 PWh. Its main advantage compared to oil is a higher energy density.

While coal gas was produced for city lights from the 1700s onwards in Europe and the USA, mining of the natural gas emerged in the late 1700s and was first distributed in tanks. The pipeline was introduced in 1821 in the USA when the Fredonia Gas Light Company delivered gas for the city lights of Baltimore.

The mining of gas was costlier than that of oil and caused explosions. Moreover, the distribution of gas stagnated because tanks were costly whilst the constructed pipelines for gas were leaking, and gas leaks also caused explosions with fires. Given the risks, the first long pipeline spanning 200 km from Indiana to Chicago was finished 70 years later, in the 1890s. Natural gas was also explosive while in use, though Bunsen burners – invented in 1885 for mixing the gas with air – enabled safer consumption of gas purchased in tanks (NaturalGas.org, 2020). Injection of chemicals for gas winning, called 'fracking', was patented in 1862. These businesses emerged in 1930 and disseminated 70 years later in a few countries because of the communities' opposition to the pollution caused by the injection of chemicals into the wells underground.

The production of natural gas stagnated for longer than one century. Losing of the lights market to electricity, failed experiments with gas heating in the early 1900s in London and other cities, and gas explosions were the major impediments until the 1960s when safer gas pipelines for heating were introduced. Rolled steel, Polyvinyl chloride (PVC) and sealing technologies enabled the construction of 1600 km of pipeline from Texas to Chicago in 1931, followed by several more pipelines in the 1950s in the USA on land and in the European North Sea. The distribution through pipes was mainly financed by the states and the production and consumption of natural gas were private. Cooling natural gas into Liquid Natural Gas (LNG) and Liquid Petroleum Gas (LPG) improved gas distribution, but these processes were costly compared to the cost of distributing them through pipes that are mainly paid for by public funds (APGA, 2020). Until the 1960s, natural gas was mainly consumed in the Haber-Bosch process of thermal reforming of natural gas into hydrogen that is used for the production of ammonia and nitrates, a process in which hydrogen is bound to nitrogen. This invention by the Badische Anilin- und Soda Fabrik (BASF) in 1913 was turned into a business in 1936 mainly due to the military interests in applications of nitrates as explosives. The production of N-based fertilisers grew in the 1960s when the purchases of fertilizers by farmers were introduced. The major impediment for growth is that the thermal reforming of natural gas is energy-intensive and polluting because per 1 kg hydrogen with 40 kWh energy-equivalent about 50 kWh energy-equivalent natural gas is needed, which causes 9 kg of CO_2 emission (Murkin & Brighting, 2016). This impediment is tackled by large subsidies for the purchases of fertilizers.

Natural gas is sometimes promoted as a backup option for the intermittence of renewable energy, but the greenhouse impact of methane emissions is seven times stronger than CO_2, and the production losses and emissions are large. The global gas leaks are about 101 billion m^3 (3.6 Tcf), equivalent to a 990 TWh loss (Forbes, 2020). In addition, about 145 billion m^3 associated gas is vented or flared in oil production, equivalent to 1400 TWh loss (Pieprzyk & Hilje, 2015). Together, these losses cover about 7% of global natural gas which causes about 8% of greenhouse emissions. This is twice higher as the global emissions of aviation but rarely addressed in the policies on climate change.

4.2.6 Hydro Energy

Continuous evaporation and precipitation on the Earth due to solar irradiation bring approximately 0.5 million km^3 water in motion; it is annually 0.5×10^{12} tonnes of water. Only 0.02% of that huge global water flow is sweet water from rivers and lakes while a small fraction of that was used for 2.7 PWh hydropower in the year 2000, which covered 2% of the global energy consumption. The production of hydropower on rivers is considered the lowest cost energy technology.

Hydropower is generated when a water flow brings motion to waterwheels as it is done in thousands of mechanical water mills for centuries. A technological breakthrough was the invention of curved blades in the waterwheels with a mechanism for steering the flows of water on the blades. Jean Victor Poncelet labelled this invention in 1838 in France as a 'turbine' (Fasol, 2002). As the energy efficiency of the turbines increased due to many modifications, hydropower was increasingly used for rotation of dynamos in generators that produced electricity. The first hydroelectric plant in 1878 in England powered one lamp, while the next one in the USA 4 years later served a few commercial customers and within one decade, a few hundred small-scale hydropower plants, which were below 25 MWe, together generated about 1 GWe capacity in the USA and Europe. Several types of turbines were developed and applied in France, Austria, England, and the USA and attuned to the local conditions. These turbines varied with respect to the volume of water, from 0.1 m^3 to a few hundred m^3 per second, and the drop of water from a few metres to a few hundred metres in height; seasonal changes are also considered in the selection and adaptations of the turbines.

During the 1900s, thousands of small-scale hydropower plants were built on the side-streams of rivers (off-river) which did not constrain the water flows. From the mid-1900s on, they were increasingly replaced by large-scale dams across rivers with barrages that created water reservoirs upstream. Such barrages (on-river) were often assumed to prevent floods downstream, to be energy-efficient and generate electricity at lower costs than off-river. As the depreciation costs of the constructions are low because the dams remain for many decades without much maintenance and are usually financed by the public expenditures, the costs per electricity unit are low. However, lower water flow, inundated communities, lost fish migration, degraded environment, and other damages are costly to societies. Dams also cause local and international conflicts about water use; for example, tensions between Egypt, Ethiopia, and Sudan about the Ethiopian Renaissance dam on Nile designed for 6 GWe (Aljazeera, 2020). For a long period, these societal costs were dismissed or neglected in the decision-making about hydropower because the societal benefits were assumed higher than the costs. These damages were increasingly acknowledged in the 1900s, as more electricity is generated in the power plants based on fossil fuel and the growth of hydropower stagnated (IHA, 2020). Whether the construction becomes more effective and cause less societal impacts is to be seen; for example, the Egyptian 2.1 GWe Aswan dam on the River Nile needs a 162 billion m^3 water reservoir while the Renaissance dam is designed for 6Gwe with half smaller reservoir, which implies 6 times improvement in the design stage.

Hydropower production from the sea also emerged. This potential is a few thousand times larger than that which is done on rivers, but the present capacity is low because construction is costly to realise and maintain. Several technologies are implemented and used. In the tidal power plants, high and low tides bring in motion rotors anchored in the sea or in barrages. In the wave power plants, energy is generated when a ball on the sea surface pulls turbines anchored on the bottom, or waves hit the turbines onshore. The ocean current power plants use rotors vested under the sea surface. While these technologies combined do not produce much power, the production grows.

4.2.7 Nuclear Energy

Nuclear energy is based on radioactive materials found on the surface of the Earth, mainly as a consequence of volcanic eruptions. The radioactive materials deliver energy when the nuclei of these materials emit neutrons, a process called 'fission'. Materials after fission, called 'isotope', are indicated by the number of neutrons; for example, the uranium isotope U_{235} is commonly used in nuclear power plants because the uranium ore is available and stable over thousands of years whereas the other isotopes are stable only several days. For electricity production in nuclear power plants, self-sustaining fission is pre-requisite. Therefore, the U_{235} ore must be purified from 0.1% to 0.7% to 4% available from mining, which is usually done by states that control this material. Enriched uranium heats up the reactor of power plants to about 300 °C for the production of steam that brings rotation in generators for electricity production. Usually, the overheating of the reactor is prevented by cooling with water (Trancik, 2006). Nuclear energy covered 2% of all energy in the year 2000, equivalent to 2.6 PWh. Its major advantage is its high energy density when compared to all other energy resources.

The nuclear power plants were invented after successful experimental self-sustaining fission in 1943, in the USA which was done for the military purposes. The first nuclear power plant started its operations in 1951 in the USA. This is followed by uses in military submarines in the USA, and by the Russian nuclear plant and military vessel within a few years. The fast introduction of nuclear energy was driven by military interests because it promised nearly infinite use of energy. The global capacity of nuclear power plants expanded from 16.5 GWe in 1970 to 135 GWe in 1980; however, the growth stagnated after explosions in the USA (Three Miles Island in 1979) and Soviet Union (Chernobyl in 1986), and came to a halt after the explosion in Japan (Fukushima in 2011). Electricity production became costly, mainly because of additional safety regulations and costly dismantling of power plants and storage of radioactive waste. These regulations also invoked the scaling up of the facilities toward a 100 billion USD investment for the 3000 MWe plant and minimum of 10 years of construction time. Frequent cost overruns in construction are experienced, though timely and cheaper cases are also observed, particularly in Korea (Lovering et al., 2016).

So far, the performance of nuclear energy has been below early expectations, despite about 100 times larger public funding for its research and development when compared to solar energy (IEA R&D, 2005). Given societal oppositions to new plants, un-insurable risks, suspicion about the proliferation of nuclear materials, and accidents, the nuclear power becomes a state asset aiming at energy independence and resource for military equipment rather than an economic resource for civil electricity generation. In addition, waste disposal is unresolved, because stable isotopes radiate for centuries (Ahaerne, 2011). A few hundred MWe power plants called 'mini' plants, thorium reactors, extracting nuclear materials from the bottom of deep seas, and other technologies are promoted, but many experts are sceptical because of past failures, high costs, and unresolved risks.

4.2.8 Geothermal Power

The geothermal energy for heating and electricity is produced from the earth's heat that is generated in its lithosphere; the lithosphere is a layer of the Earth found at a depth of 400 km from its surface. Traditionally, geothermal energy is made available as a result of geyser and volcanic activities on the Earth's surface. Industrial production of geothermal energy also uses heat mined in the ground and from groundwater. Such mining requires drilling in the areas of high-density heat under the Earth's surface. Therefore, the geothermal maps show heat areas to be explored while oil drilling equipment is used for this purpose. The share of geothermal energy grew slowly from the early 1900s onward toward 0.2% of the global energy production; it was equivalent to 0.18 PWh in the year 2000. Its main advantages among low-carbon energy technologies are continuous delivery of heat and power and the use of a small surface when compared to wind and solar energy, as well as less contentious applications compared to hydro and nuclear energy.

The first industrial geothermal plant for heating was successfully constructed and used in Italy in 1904, followed by a few heating plants within several years, operating in small scale industries. The electricity generation in power plants emerged 60 years later in the USA, Japan, and New Zealand without commercial successes during its operations. Their energy efficiency rarely exceeded 10% because of insufficient heat for steam production that is needed for the power generators whereas the delivery and sales of heat did not cover the costs. As a result, most power plants in the USA and Europe are demonstration plants. Only a few full-scale geothermal power plants that deliver electricity and heat are vested in areas of volcanic activity such as Costa Rica, Iceland, Indonesia, the Philippines, and the USA.

The major impediments for the dissemination of geothermal power are costly drilling of boreholes and corrosion of the equipment. A geothermal heater needs a borehole that is a few hundred metres deep and a heat exchanger; it is a device that transfers heat from the borehole to the heater, and vice versa, without transferring materials from the borehole. In this manner, the heat exchangers prevent the entry of corrosive borehole materials into the heating system on the surface, while the heat exchangers corrode if located without protection underground. The construction of a geothermal electricity plant is costlier than heat plants because of deeper drilling and costlier equipment; a borehole that is a few kilometres deep is needed. Therefore, equipment must withstand temperatures of a few hundred centigrade Celsius and pressures of a few hundred bars. A heat exchanger in the borehole is needed along with a powerhouse on the surface, which enables to generate steam and pressure for the motion of the generator. In the powerhouse, costly organic materials are used which support the generation of high pressure for the motion of the generator without very high temperature.

Small-scale geothermal heat transfer of underground water for housing is more successful. They were introduced during the 1940s and grew in the 1980s when the energy prices went high. These devices use heat transfer from underground water and combine it with heat pumps, which are electric devices that increase the temperature of the heat exchangers with use of electricity. Given the temperature stability of underground water, hot water used in households is cooled underground during summers; and cooler underground water is pumped and warmed up with heat pumps for heating during winters. These boreholes are a few dozen metres to a hundred metres deep. Despite the huge potential of geothermal heat, its application progresses slowly because of low energy efficiency.

4.2.9 Wind Energy

Wind energy is produced due to air movement in the atmosphere generates vertical and horizontal pressures called ‘lift’ and ‘drag’. These wind pressures on sails bring ships in motion, and on blades bring the axis of a wind turbine in rotation, which generates electricity. The rotation needs wind speeds between 3 and 20 metres per second; it is because lower wind speeds provide insufficient power for the rotation, while stronger winds can damage the turbine. In theory, wind turbines can use nearly 60% of the wind power, but they usually attain not more than 50% conversion; this is similar to fossil fuel power plants. The share of wind power increased fast from near nil in the 1980s to 0.03% of the global energy production in the year 2000, equivalent to 0.03 PWh. Its main advantage when compared to other energy technologies is its wide availability.

In addition to sails on ships for motion, mechanical windmills are used for transport of water and loads, milling of minerals, grains and such like heavy works. Thousands of mechanical windmills with the horizontal axis were used for centuries; these are ones with vertical blades. The first horizontal axis turbine for electricity was constructed by James Blyth in Scotland, in 1887. It delivered electricity locally; however, the installation was dismissed by the local community as the ‘work of the devil’. A larger, 17 kWe turbine was constructed by Charles Brush one year later in the USA. It delivered a direct current (DC) on his farm for over 20 years. Many small-scale units followed; for example, the Jacobs Wind Company sold about 30,000 units worldwide between the years 1922 and 1950. During that time in Europe, Denmark was leading with a total of 30 MWe capacity from 2500 units (Dismukes et al., 2009). The first large-scale wind turbine was constructed by the students and teachers on the Danish Tvind School in 1978. These tinkerers generated power from the late-1970s on. Their open-source design developed by the users of equipment was improved upon by the producers as Vestas, Siemens and other firms in the mid-1980s (Shahan, 2020) which illustrates the scaling up of the users' innovations. After the 1980s, the scale of wind turbines doubled every second year measured by power capacity which reduced costs per MWe. The turbines' capacities increased from 1 MWe to 5 MWe per unit, whilst their rotor diameters also enlarged from 20 to 110 m by the year 2000; the latter is larger than the largest airplane wings. Policies in Denmark, Germany, Spain, and other countries in Europe encouraged this development with standards for quality performance and subsidies per capacity (Bergkek & Jacobsson, 2003). The price parity with electricity on-grid is reached by the wind turbines on land, if intermittent delivery is ignored.

Meanwhile, those horizontal axis turbines expand in shallow water, called 'offshore wind'. Despite costlier construction than the construction of 'onshore wind', the building processes are faster because strict regulations on land are not valid on seas and energy production is higher due to more continuous and stronger winds. Winning tenders for offshore projects also produce power at a similar price as coal power plants. After the year 2000, wind production grew fast in several European countries due to high fuel prices and price guarantees for deliveries of renewable energy to the grid. These price guarantees called 'feed-in tariffs' for wind power created a stable market; meanwhile, in the USA, producers obtained subsidies which reduced costs but did not foster stable market.

The vertical axis turbines, ones with horizontal blades, were invented by Jean Marie Darrieus in 1931 in France. They are integrated on some buildings, ships and operate in niche markets. This design of wind turbines lost the competition with horizontal axis turbines as they were less energy-efficient. Kites, sails, vibrating columns, and other novelties in electricity generation from wind energy emerged, but they were so far commercially unsuccessful. Floating large-scale turbines offshore on open seas with horizontal or vertical blades also emerged in the 2000s.

4.2.10 Solar Energy

Solar energy is produced through concentration of the solar irradiation, for heating and electricity. The production encompasses solar heaters, concentrated solar panels (CSP), and photovoltaic panels (PV), which employ different concepts for capturing irradiation and different patterns in the technology development without many crossovers. Although the total consumption of solar energy is low because it held only a 0.001% share in the year 2000, equivalent to 0.0012 PWh, expectations for the future are high, and the growth rate is in the double digits. Its major advantage is the availability of solar irradiation across the world during daylight.

Solar heaters emerged as black coloured water tanks with insulation in the 1800s. After re-designs in 1891, by Clarence Kemp in the USA, they evolved into solar boilers for hot water during the 1920s in the USA. While solar boilers were widely used for cheap water heating throughout the 1900s in the USA and Israel, not many countries followed this suit. After the 1970s, solar boilers were largely replaced by solar collectors. Most simple solar collectors have coated metal plates with an insulated tank for water storage. Thereafter, effective air-based solar collectors were developed. More effective collectors were introduced based on vacuum tubes with fluid organic chemicals that transfer the irradiation into heat storage, but storage of heat during several weeks remained challenging. These vacuum tubes emerged after the 1980s. They disseminated slowly because tanks for heat storage are insufficiently effective during cold periods.

Concentrated Solar Panels (CSPs) concentrate the irradiation in hot air, steam, or fluids, which bring the turbine in motion. The first working CSP-based steam engine was invented by Auguste Mouchout in 1866 in France, and the working CSP for electric power was constructed by Frank Schuman in 1912 in Egypt. This was nearly 70 years before the first modern CSP was constructed in the USA. After the 1980s, two rival systems emerged and disseminated. In one system, the concentration of irradiation was achieved with parabolic mirrors wherein organic fluid at the bottom of the mirrors is heated up and transported to a power plant; an example is a power plant in the Mojave Desert in the USA. In another system, lenses are focused on heat receivers in a tower where heat is concentrated in the organic fluid that is transported to a power plant; an example of this was constructed in Spain. The dissemination of CSP was slow because operations were costlier than the large-scale thermal electricity generation, except in areas with very high irradiation; for example, the price parity in distribution on-grid was reached in the winning tenders in Abu Dhabi and Morocco.

Most solar power is produced with photovoltaic equipment (PV). First PV was invented by Charles Fritts, who installed a selenium coated panel that produced a weak electricity stream in 1883 in New York; this is an illustrative example of technological breakthroughs through tinkering because this PV installation was operating 20 years before physics had gained an understanding of the links between light and electricity. However, this invention did not maturate in a commercial business. Seventy years later, Bell Laboratories in the USA patented a PV cell in 1954, which was achieved thanks to military funding and interests for applications on airplanes and satellites; Sharp in Japan, Siemens in Germany, Philips in the Netherlands, and other firms worked on similar cells from the 1950s onwards. By the 2000s, the Chinese companies became largest producers of PV and the PV-based solar panels.

A basic element in the PV is the photovoltaic cell. On the front side of the cell, the side directed towards the Sun, the light elements (photons) are captured using light-sensitive materials which trigger the release of electrons. Within the cell, the electrons released are conducted through the electricity-sensitive material to the back-side of the cell, where the direct current is generated. Most cells are constructed with a silicon material, called 'crystalline PV', while other semiconducting materials are also used for their high energy efficiency, flexibility, and other properties. A large number of cells are composed into a unit block, called 'wafer'. A large number of wafers constitute a photovoltaic module with connectors to a grid, called the 'PV panel' or 'solar panel'. A panel delivers a direct current which can be transformed into the alternating current with a transformer. PV applications emerged in shuttles, telecommunication, gadgets, and other niche markets in the 1970s whenever the independence of a grid provided advantages to users. PV was disseminated into mass consumption due to the standardisation of panels during the 1990s. Several panels constitute power plants, from 5 kWe in households to 50 MWe for businesses. PV grew fast during the 2000s. As the production volume of panels increased, the costs of solar energy declined fast and approached price parity on the electricity grid in the 2010s in the countries with high solar irradiation. The feed-in tariffs for solar power enhanced that growth.

4.2.11 Timeline

This overview shows highlights from the history of 29 basic energy technologies for 10 resources introduced from 1800 to 2000 and is focused on the lead times from inventions through innovations, towards diffusions. Those technologies emerged in the last 200 years and nearly all are still used after many modifications, which underpins complementation mechanisms in economies, rather than substitutions. The diversification of energy technologies is observed throughout two centuries, rather than optimisation toward the lowest unit cost technology as assumed in the neoclassic theory, or evolution in the direction of technological lock-in as argued in evolutionary economics.

Patterns in technological change vary in time, measured by the number of technologies, periods of innovating and growth rates. The applications of all those technologies are adapted to local physical and socio-economic conditions during those centuries. Aside of such adaptation, the basic energy technologies diversified. There is variation by the number of technologies. While the biomass, coal, electric, oil, hydro, gas, and solar resources are diversified into several technologies that operate side by side, the nuclear, geothermal, and wind resources evolved toward a few designs per technology with various adaptations. For instance, one finds five basic designs of turbines for conversion of the water resource into hydropower, but two basic designs in wind turbines (horizontal or vertical rotors). There is also variation per period. Most fossil fuel technologies grew between 1870 and 1920 and renewable energy technologies expanded between 1980 and 2010, whereas only a few technologies grew before or after those periods of global income growth, which suggests that the upswing in income growth triggers the introduction and dissemination of innovative energy technologies rather than the economic stagnation and so called degrowth. However, the patterns of growth also vary. Whilst the basic technologies in bioenergy, coal, hydro, and geothermal energy disseminated steadily for decades, ones in oil, gas, nuclear, wind and solar expanded within a short period of time. These variations imply that forecasting technological changes is difficult, if possible. They also indicate that 'lock-in', 'transitions', 'landscapes', and other catchy metaphors in the modelling of technological changes can cover up an insufficient understanding of innovation processes.

The lead times cover several decades and vary across the energy technologies. Those lead times divided into the development phase from inventions to innovations, and the diffusion phase from innovations to disseminations are presented in Table 4.2. Within those 29 basic energy technologies, 19 technologies had lead times between six and nine decades which can be considered as a typical time span from inventions, through the start-up of innovating firms and survival in the competition. These typical lead-times of innovation processes refer to biodiesel large steam engines, coal-based boilers, fluidized bed combustion, batteries for direct current, rechargeable batteries, alternating current, oil wells, four-stroke engines for gasoline, kerosene for aviation, gas heating thermal reforming, hydropower turbines, geothermal power plants, heat pumps, horizontal wind turbines, vertical wind turbines, solar boilers, and concentrated solar power. Among those innovation processes, the development phase and diffusion phase were three to five decades each, usually

Table 4.2 Lead times of basic energy technologies measured in decades from 1800 to 2000 divided into the development phase from inventions to innovations, and the diffusion phase from innovation to dissemination, based on author's estimates

Resource	Technologies	Development	Diffusion	Lead-time
Bio	Industrial bio-alcohol	6	4	10
	Digestion for biogas	3	8	11
	Biodiesel production	1	5	6
Coal	Large steam engines (earlier)	5	2	7
	Coal based boilers	4	3	7
	Fluidised bed combustion	4	2	6
Electro-chemical	Batteries for direct current	5	4	9
	Rechargeable batteries	3	5	8
	Electricity, alternating current	5	4	9
	Fuel cells	9	3	12
	Laser	1	2	3
Oil	Tar sands oil production	4	8	12
	Oil wells production	2	4	6
	4 stroke engines for gasoline	2	4	6
	Diesel for fast movers	1	3	4
	Refined kerosene for aviation	7	2	9
Gas	Pipelines distribution	10	3	13
	Gas heating	2	6	8
	Thermal reforming	2	4	6
	Fracking	7	6	13
Hydro	Hydropower turbines	2	4	6
Nuclear	Nuclear power plants	1	2	3
Geo-thermal	Geothermal power plants	4	4	8
	Heat transfer and pumps	4	2	6
Wind	Horizontal rotor	2	4	6
	Vertical rotor	5	2	7
Solar	Heaters and boilers	3	5	8
	Concentrated Solar Power	7	2	9
	Photovoltaic	1	3	4

with several new designs, many adaptations and applications. Only four technologies expanded within six decades. All of them were driven by the military interests of states, which were diesel for the British Navy in the 1910s, nuclear plants for ships and submarines of the USA and the Soviet Union in the 1950s, photovoltaics for telecommunication, and space shuttles in the USA and the Soviet Union, and laser technologies against missiles for the so called "Star Wars" of the USA in the 1980s. Contrary to these short lead-times, the lead-times of six technologies covered ten or more decades. These were industrial bio-alcohol, biogas, hydrogen with fuel cells, tar sand for oils, gas pipelines, and fracking of gas. All these innovation processes faced resistance to implementation on account of societal concerns about explosion and pollution.

This assessment indicates that most lead times are far longer than one generation, which implies that thorium reactors, deep geothermal mining, deep-sea nuclear power, geoengineering of the irradiation, photo-bioelectric technologies and several envisioned technologies for energy storage cannot contribute to the mitigation of climate change within one generation and can be useful in a long run if low-impact. Devoting much attention and allocating large funds for the imaginary technologies is risky. It is even irrational if severe impacts can be expected because social concerns about technological hazards slow-down the diffusion of novelties for decades. Particularly high-density energy resources invoked societal resistance with regard to hazards. For example, explosions of natural gas obstructed its scale-up for nearly one century, hydrogen applications in mobility were stopped for 80 years after the explosion of a hydrogen-driven zeppelin "Hindeburg" at the docking station in USA in 1937, nuclear power promised safe, infinite energy, but failed after two decades because of accidents during the 1980s and did not revive until after 40 years.

Technology assessments are instrumental when used to assess possibilities and constraints in the future. Yet the technology forecasting can create misleading images about solutions for pressing problems. For example, the image of Haldane in 1923 was about "great power stations where during windy weather the surplus power will be used for the electrolytic decomposition of water into oxygen and hydrogen." (Haldane, 1923). However, this idea is blown up by that catastrophe in the docking station 14 years later. Hydrogen was used in industries, but applications in mobility emerged 60 years later by the early-2000s based on state funding because it was experienced that hydrogen is costly and explosive. The 'Flinstones family' in comics of the 1960s drove nuclear cars but such images of nuclear applications vanished within a few decades because radioactive hazards are experienced in power plants. A dose of technological optimism is needed for innovations. Nevertheless, introductions of novelties require societal debates in order to avoid human casualties, high costs during diffusion, and long delays caused by risks. It should also be acknowledged that most people tend to avoid risks. As incomes grow safety, well-being and healthy environments are increasingly valued (Thurow, 1980).

The maturation of basic energy technologies takes usually six to nine decades and longer when safety and environment are at stake; and when energy technologies are introduced within a short period of time their diffusion stagnates if they turn to be risky and polluting. Those long lead times imply that greater access to energy and mitigation of climate change needs to be attained with new designs in the available energy technologies whilst the development of new basic technologies can be pursued along with the impact assessments.

4.3 Business Interests in Innovations

Given the long lead-times of innovation processes in the basic energy technologies, it is assessed what innovations in energy catch high business interests In line with the mainstream argumentation, it is assumed that the business interests are high if expenditures in research and development (R&D) generate many patents. High R&D expenditures indicate priorities for specific technologies whereas many patents express business initiatives; however, many firms operate successfully without patents, and many patent holders do not start a firm. For answering that question, the R&D expenditures and patents in all sectors of business, the energy and renewable energy are compared. The global R&D expenditures and the number of patents are estimated, as well as their distribution across 14 countries with more than 100 million inhabitants for indication of the business interests across many countries. A high share of R&D expenditures in renewable energy, when compared to energy R&D and all R&D indicates the priorities in technology development, and many patents in renewable energy indicate high business interests.

A brief explanation of the data is useful. Data are based on the statistics of the World Bank, IEA, and OECD (OECD RD&D, 2021), whereas the missing data are interpolated or extrapolated. R&D expenditures in all businesses are based on the World Bank database. For Nigeria and Bangladesh, they are proportional to scientific publication because their R&D data are scarce. IEA data is used for the energy R&D. These are divided into 8 objectives in the IEA database: 'energy efficiency', 'fossil fuels', 'renewable energy', 'nuclear energy', 'hydrogen and fuel cells', 'other power and storage technologies', 'another cross-cutting technologies/research', 'unallocated', and total expenditures (IEA RD&D, 2021). However, many data on the EU 28 member countries are interpolated. This database shows global energy R&D, as well as that of the USA, Japan, EU, and Mexico. The energy R&D in other countries is estimated in the following manner: the energy R&D of those four countries is subtracted from the global energy R&D and divided across the remaining ten countries in proportion to their entire R&D expenditures. This is merely an approximation of the countries' energy and renewable energy R&D. The OECD data on patents show the patent families (IPS), which are clusters of related patents because inventions are often covered by several individual patents when aiming to accrue monopoly in a particular technology. Herewith, the patent families in electricity technologies indicate the patents in all energy and ones in the mitigation of climate change illustrate the patents in renewable energy; both refer to energy generation, transmission, or distribution. In addition, individual patents are used, which are categorised as 'renewable energy resources' and 'energy generation from non-fossil origin' in the patent group on climate change mitigation within the environment-related technologies.

Global expenditures on the energy R&D are in line with global expenditures on energy production while there are differences with respect to R&D of particular energy resources. While globally, all R&D covered annual average expenditures of about USD_{2018} 1500 billion during 2008–2017, which was 2.1% of global GDP, about 1.9% of these expenditures were allocated in the energy R&D which is about as much as the costs of energy in the GDP. Within the energy R&D, the expenditures on nuclear power are excessively high but ineffective because 25% of global energy R&D on nuclear energy covers provides only 2% of the global energy production. Meanwhile, the R&D on fossil fuels was low because only 13% of global energy R&D while fossil fuels covered about 80% of the global energy production, excluding nuclear power. Renewable energy R&D is reasonably high because it covered 19% of energy R&D compared to 14% share in energy production.

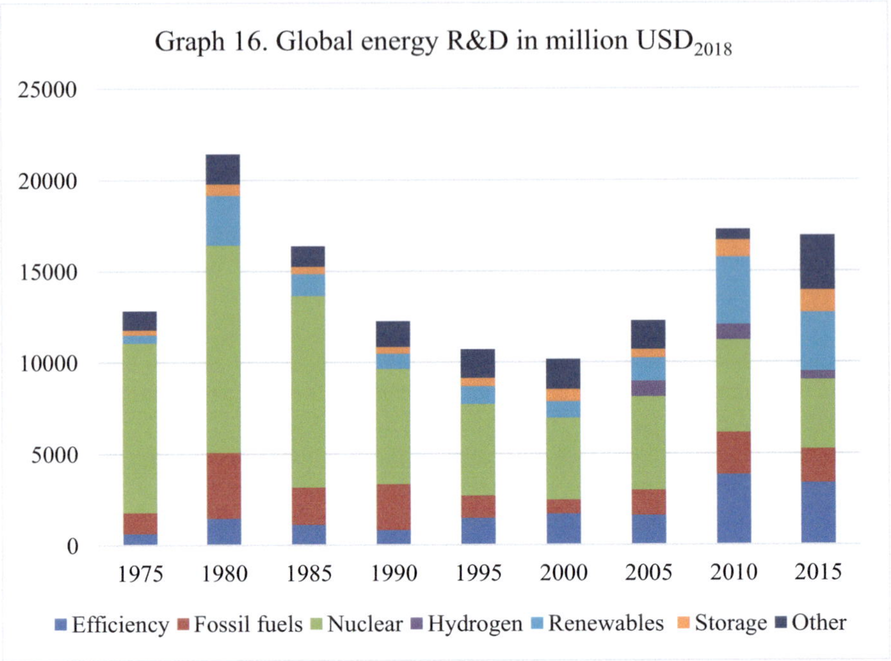

Fig. 4.1 Global energy R&D in million USD_{2018}

Large differences across countries are shown in the World Bank and IEA data. The energy R&D was concentrated. The USA, Japan, and the EU covered 84% of the global expenditures, whilst a similar share is observed in renewable energy R&D. All R&D of most other countries is low compared to GDP, energy R&D in all R&D is also low, but renewable energy R&D is energy R&D is close to nil. For example, while the GDP-PPP while GDP-PPP of low-income countries is up to 25 time lower than of Japan, their energy R&D is up to 1000 times lower and renewable energy R&D up to 10,000 times lower. Renewable energy R&D is close to nil in most low-income and mid-income countries, and high in high-income ones, measured in total expenditures and share in all R&D. R&D expenditures in most mid-income and low-income countries address mainly fossil fuels. These countries focus on development of fossil fuels, though they have signed the international agreements on the mitigation of climate change.

Changes in the objectives of energy R&D over time indicate factors that drive priorities in the energy business. Based on the IEA data, Fig. 4.1 shows the R&D expenditures in USD_{2018} from the years 1975 to 2015 in 5-year intervals and divided into six categories: energy efficiency, fossil fuels, nuclear and fuel cells (hydrogen), renewable energy, other power with storage technologies (storage), and cross-cutting and other unallocated ('others'). The latter two categories mainly refer to technologies that link energy to information and communication technologies (ICT) in the distributed energy systems.

All global R&D expenditures steadily grew based on the OECD with World Bank data but the energy R&D, hardly increased between 1975 and 2017 which indicates low business interests in innovations in energy. Throughout those 40 years, the global energy R&D in USD_{2018} grew at an annual average of only 1%, which is below the income growth. The renewable energy R&D grew 6% per year. Meanwhile, R&D in the renewable energy system, which includes storage and ICT, grew a few percent faster. The interests for innovations increased particularly for renewable energy combined into the distributed energy systems, less so in other areas of the energy R&D.

The energy R&D fluctuated in line with the international prices of fossil fuels. It increased during high prices in the 1980s, and decreased by half during low prices; then, increased again from the year 2005 to 2015, and declined thereafter. Within the energy R&D, the R&D expenditures in energy efficiency and renewable energy fluctuated even more because they increased faster toward a larger share in energy R&D when the prices went up and declined faster when the prices went down. In particular, the expenditures on renewable energy R&D increased sixfold to 16% of all R&D during 1975–1980; then, declined to one-third of it during the low prices of the 1990s; thereafter, increased fourfold from 2000 to 2010, when they covered 27% of the energy R&D; followed by a decline to one-fifth of the energy R&D that also declined. Those fluctuations show that the renewable energy R&D is strongly influenced by the market prices as assumed in mainstream economics, which also implies that the policies on energy pricing generate business interests for innovations in renewable energy, whilst the policies that support fossil fuels just as effectively kill these interests.

Meanwhile, the R&D expenditures into nuclear power, hydrogen, power with energy storage, and 'others' steadily changed, but the directions of these changes differed. The R&D into nuclear power covered about half of total energy R&D from 1975 to 2017 – cumulatively USD_{2018} 290 billion – but declined steadily from 75% to 21% in the energy R&D. This illustrates the declining business interests along with a vocal business and science lobbies for larger investments. During that time, the R&D in power with storage and 'others' increased fourfold, which shows the growing business interests in the distributed energy systems. The R&D into hydrogen emerged after the year 2000, but its growth saturated after 10 years and declined thereafter which indicates a declining business interest despite loud calls for more public expenditures in this technology. In sum, the R&D from the late 1970s to the late 2010s shows growing business interests for innovations in the demand-oriented renewable energy on-grid and off-grid, it is the growing business interest in the renewable energy systems (RES) whilst all other areas of energy R&D are perceived as minor opportunities.

Regarding the popularised ideas about mission-oriented policies, an issue is whether policies can be effective in the development of energy technologies. Experiences of the past decades with the public funding of the energy R&D show mixed results, at best. Policies were successful in lasers and photovoltaics, but manifold larger public R&D expenditures in nuclear power failed regarding its low and stagnant global share in energy production. Even larger public R&D expenditures are devoted to nuclear fusion; for example, solely a demonstration plant for nuclear fusion of USD

22 billion in 2021 (Gibney, 2021) is costlier than a full-scale nuclear power plant. So far, the nuclear fusions failed and those expenditures cast doubts on if its full-scale production can ever cover costs given the rivalry of cheaper renewable energy. Large public R&D expenditures are also allocated into hydrogen production with electrolysis but commercial successes are not attained despite two centuries of experimentation. The development of the fuels cells for hydrogen production took one and a half centuries without commercial successes, whilst the batteries became more energy-efficient and cheaper rivals to fuel cells. Last decades, the business interests in the future of fuel cells weakened regarding the declined R&D. Meanwhile, large subsidies for bulky hydrogen production with electrolysis are announced by policies in the EU and Japan. Regarding the declined business interests, it is presumably useful for basic, energy-intensive industries but a costly failure for other consumers in business and households compared to expansion of batteries.

While markets punish high-risk investments through competition for capital – even if far from perfect – corrective mechanisms for the public R&D are intangible checks and balances, at present. Many failures in the mission-oriented energy R&D policies were experienced because decision-makers in policies pursued particular technologies when motivated by the particular businesses rather than posing far reaching demands aiming to resolve pressing societal issues through valuable energy services. Therefore, policies pushed technologies rather than created demands for innovations. When the mission-oriented policies push technologies, they open the box of policy failures as firms, consultants, scientists, and policymakers become addicted to subsidies for a particular technology entailing lock-in into the particular private interests. Contrary to that, posing far reaching demands for improvements creates opportunities for the entrepreneurial initiatives that generate innovations which diversify technologies.

The results of R&D are measured by the number of patents. They indicate business interests because patents are assumed to support the start-up of firms. All patents are compared to the energy patents, and to the renewable energy patents globally, and across countries. This is followed by more detailed estimates of the specific patents in renewable energy. Figure 4.2 shows the patent families in renewable energy, including biofuels, across countries from 1990 to 2017, using OECD data on the patent families in the technologies for the mitigation of climate change.

All patent families grew by an annual average of 2.5% from the year 1990 to 2017, whilst the energy patent families by 3.8%, and the renewable energy ones by 5.1%, which means three times, four times, and seven times more patent families in 2015 than in 1990, respectively. Note that the energy R&D per patent declined given the slower growth of the energy R&D than the patents. Effectively, the share of the energy patent families in all patents increased from 17% in the early 1990s to 26% in the mid-2010s, and the renewable energy patent families among the energy families grew from 5% to 8%. This means that renewable energy generated larger business interests compared to other sectors and that the renewable energy R&D was increasingly effective, measured by the number of patents. In sum, the business interests in renewable energy grew and the continuation should be expected unless the global policies would discourage this trend.

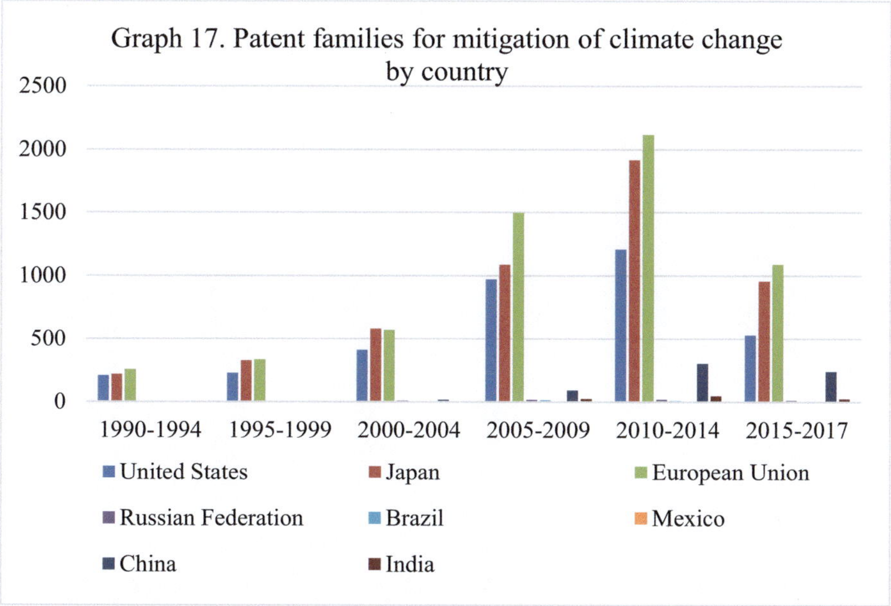

Fig. 4.2 Patent families for mitigation of climate change by country

Across countries, access to novelties in energy and renewable energy improved. While the USA, Japan, and the EU together obtained 92% of all patent families in energy and 91% in renewable energy in the early 1990s, these shares declined to 61% and 68% in the 2010s. Particularly Korea, China, India, and Russia gained patents in energy, whilst the patents in renewable energy grew in more countries. Nevertheless, the EU holds about a third of the latter. However, business interests in energy and renewable energy declined after 2010, which can be attributed to declining fuel prices and the deficient data, as the approval of patents and processing into the statistics take several years.

The R&D of energy and renewable energy were also efficiencies measured by low R&D costs per patent. On average between the years 1990 and 2017, the unit cost of energy R&D per energy patent was about USD_{2018} 0.3 million. The unit costs were similar across most energy R&D objectives: USD_{2018} 0.1 million per patent in storage and distributed energy systems, USD_{2018} 0.2 million in renewable energy and hydrogen, USD_{2018} 0.6 million in fossil fuels; however, USD_{2018} 7.6 million in nuclear power. These unit costs did not vary much across the high-income countries whilst they varied greatly across mid- and low-income countries. The span of this range varied from more efficient R&D than high-income countries in the Philippines and India to 10 times less efficient R&D in Nigeria and Brazil. Globally, the efficiency-increase of energy R&D was about a 4% annual average during that

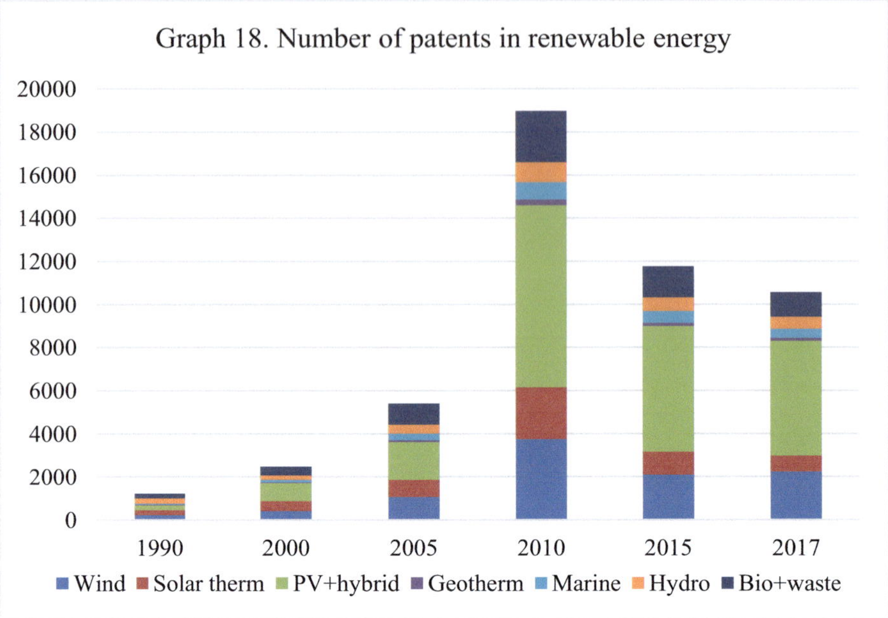

Fig. 4.3 Number of patents in renewable energy

period, which was similar across all objectives in energy and did not vary across high-income countries, though it did vary across mid-income and low-income countries. This efficiency-increase in the R&D indicates that the allocation of energy and renewable energy R&D to countries with efficient R&D generates large business interests, and it is a successful innovation strategy.

Those patent families in the mitigation of climate change cover individual patents in renewable energy, which indicate more specific business opportunities. Figure 4.3 shows the number of patents in wind, solar-thermal, solar PV including solar-thermal hybrids, geothermal, marine, hydro, and bioenergy with biowaste. They are shown in 5-year intervals from 1990 to 2017; year 2017 is not proportional on the horizontal axis.

While in the year 1990, the number of patents for the production of renewable energy including biofuels was hardly 0.3% of all patents for the mitigation of climate change, their share increased to 2.7% in 2010; most patents addressed energy efficiency of the industrial production. A particularly fast increase was in PV, including hybrid thermal-PV. As a result, the patents for PV within all renewable energy patents covered 18% in the year 1990, and 50% after 2010. The growing share underpins the growing business interests in solar energy. Meanwhile, the number of patents in wind energy also increased but its share declined, indicating that the growth of these business interests gradually saturates. The number of patents in other renewable energy technologies declined fast after peaking in the year 2010, despite high fuel prices until 2015.

While markets and policies failed in the R&D in nuclear power and hydrogen, measured by the number of patents, the renewable energy R&D was more effective. Business interests in renewable energy grow particularly fast in PV linked to information and communication for the demand-oriented renewable energy systems on-grid and off-grid. The R&D in renewable energy in high-income countries is effective, and it improves fast in a few mid- and low-income countries which implies that allocation of R&D to the countries with an effective R&D generates opportunities for innovations.

4.4 Chances for Innovations

Regarding large business interests in renewable energy, an issue is if particular R&D proposal can be assessed as potentially successful innovations in the future, given that the events of the future are uncertain. A usual method is assessing of the inventor's track records with respect to past performance, exceptional capabilities and other proven qualities. However, it is observed that firms without long references and new managers in the vested firms generate more successful innovations than the incumbents, because the newcomers act at their best and can avoid vested routines (Kaplan et al., 2009). Another option is awarding of the R&D proposal assessed by several experts with respect to scientific qualities, impacts, costs and other criteria whereas verification of the assessment can only be done after the realisation and use of novelty. Meanwhile, the experts' opinions form the bread and butter in policies aiming to grasp the uncertain future. While high profile committees with large expertise are often established to assess the best proposal and models with 'big data' are popularised (Bonaventura et al., 2015), these assessments are guesses biased by the selections of experts and interpretations of their opinions. In effect, risky R&D proposals are often dismissed and eventually become successful, and less risky ones are assessed as promising and fail as they add little value compared to solutions known from the past (Perrin, 2021). Such flaws in the decision-making about R&D can partly explain the emergence of unexpected novelties that replace the incumbent technologies called 'disruptive innovation'.

Flaws in the decision-making based on the expert opinions are illustrated in the evaluation 'Horizon 2020' which has been the main EU program for the R&D grants between 2014 and 2020, one of the largest public funding of R&D in the world. The R&D proposals aiming to resolve major societal challenges, which were assessed as excellent, are awarded by the European Commission, the executive body of the EU; given focus on the societal challenges as aging, energy, climate, and suchlike, this public funding can be considered an example of the mission-oriented policy (EC, 2021). Each R&D proposal of a cross-countries consortium of scientists, businesses, and social organisations from within and outside the EU is assessed by a few experts independent of each other with respect to the proposed content and impact, as well as the capabilities of the consortia. The evaluation of 'Horizon 2020' shows that 115,235 eligible R&D proposals for €182 (USD 217) billion were submitted during those 7 years, while 13,903 grants were approved for €25 (USD 29) billion, which generated

141 valid patents. Thus, the chance of approval was 12%, for 14% of the budget, and one in about hundred projects that were granted delivered a patent; the latter makes it 1.01% chance that the approved R&D project provides a patent at €0.18 (USD_{2010} 0.21) million per patent (Horizon2020, 2021). If the chance for approval of the proposal is combined with the chance that the project has generated a patent, one in a thousand of R&D proposals has generated results that provide opportunity for an innovation. One percent success of approved R&D projects means high rate of failures by the experts in assessing of the R&D proposals despite strict selection criteria and comprehensive selection methods Presumably, the experts can assess the internal consistency of proposals, but not more than that. This is neither encouraging for the participants in consortia nor for resolving major societal challenges, taking into consideration that the approved patent is only a stepping stone in innovation and diffusion. Given that the EU program generates excellence in science and technology, subsidies for a particular R&D are doomed to fail if it is aimed to generate an innovative start-up.

Another approach aims to foster conditions for the innovating process; meaning creating the higher chances for the expected income. Learning is that projects with highest expected income should be selected, meaning the highest Net Present Value of cash flows after corrections for the probabilities of costs and income in the innovation process (Sullivan et al., 2009). In this reasoning, the probabilities are interpreted as chances of success and the chances are assumed to be externally given. However, this method does not help much for if the chances are low – for example, as low as mentioned above – the expected income is about nil. Even worse, the combined chances of the approved R&D proposal and patenting, compounded with the chances that the patents generate sufficient income for start-ups to survive in the competition could be lower than gambling in the casino roulette. If it is aimed at higher chances of innovations, better conditions for successful R&D need to be assessed and realised. This possibility to assess and realise better conditions for the innovation processes across countries is discussed below. For this purpose, it is assumed that the R&D ideas aim to generate patents, patents aim to start a firm, and startups pursue survival in competition. However, in the innovation process, the R&D can fail to provide patents, the patents may not be used, or firms can start without patents, and the startups can fail in the competition.

Therefore, statistical data on the R&D expenditures, numbers of patents, the firms' births, and surviving firms in the period 2008–2017 are used. The births of firms indicate the startups, and the surviving firms are estimated based on the yearly increase minus decrease of the firms. The estimates that address all sectors are called 'ALL', and ones that cover renewable energy in the broad sense are called 'RES+'. RES+ refers to the energy R&D excluding fossil fuels, which covers patents in energy efficiency, renewable energy, nuclear energy, hydrogen, and storage, and to the firms' birth and survivors in energy. That broad coverage is used because more specific data is not found, whereas the startups are usually within those areas. ALL data are based on the World Bank statistics, and the survivors' data is found in the countries' statistical yearbooks. For the RES+, the R&D and patents data are based on the IEA and OECD statistics, but the firms' data is found only in the USA Census and the EU Eurostat, not in the statistical yearbooks. These estimates cover the period 2008–2017 when the conditions for innovations in renewable energy were

Table 4.3 Countries R&D expenditures, patents, firms' birth and firms that survived during 2008–2017 in ALL sector and RES+ based on IEA, OECD and World Bank data

R&D is in USD_{2010} million	R&D per PATENT A	Patents per firms' birth B	Firms birth per survivor C	R&D per survivor $D = A \times B \times C$
All sectors				
United States	0.8	0.8	26	17
Japan	0.5	23.1	−0.2	−2
European Union	2.7	0.1	6	1
Russia	0.4	0.1	−12	−1
China	0.2	3.9	48	41
Korea Republic	0.3	2.5	1	1
RES+ sectors				
United States	5.9	1.2	3	23
Japan	1.8			
European Union	3.1	0.1	1	0.37
Russia	12.9			
China	5.1			
Korea	0.8			

favourable due to the economic recovery after the financial crisis and high international prices of fossil fuels.

Those innovation processes in the EU are compared to the USA, Japan, Russia, China, and Korea. This is followed by the estimate of the rate of success in innovations processes in the EU member countries. These rates in the EU countries are estimated in two steps: a number of patents (R&D output) per number of R&D projects (R&D input) called 'development performance', and a number of survivors in competition (firms' output) per firms' births (firms input), labelled 'diffusion performance'. Multiplications of the development performance and diffusion performance indicate the rate of successful innovation processes, called 'innovation performance'. This is indicative because R&D results are not necessarily expressed in patents, or other IPR and many startups do not use patents; nevertheless, technologies are often patented. As the EU countries vary in population, from less than half a million to more than 80 million people, and their income from high income down to income below USD_{2010} 20000 per capita, those estimates of innovation performance are illustrative for many other countries in the world.

The R&D expenditures per survivor are based on multiplications of USD_{2010} per patent, times patents per firm birth, times birth per survivor in the ALL and RES+ sectors. The results are shown in Table 4.3; recall that the data on RES+ are found only for the USA and EU. It is an acceptable omission because these countries together encompass most of the global innovation processes as they cover about 75% of the global RES+ R&D expenditures, 98% of the RES+ patents, and 82% of the RES+ firms' birth.

The results are discussed in comparison with the EU. The R&D expenditures per patent in ALL sectors are usually below one million dollars, except in the EU where they are USD_{2010} 2.7 million per patent. The latter is not because R&D in the EU is costly. The EU inventors are not as patent-driven as those in the Asian countries, and the USA. For example, a European startup uses 500 times fewer patents than a Japanese one. Nevertheless, the startups survive more often in the EU than in other countries, nearly as often as in Korea. In effect, the EU innovation performance in ALL sectors is higher than in other countries, whilst similar to Korea. In the RES+ sector, the EU R&D expenditures per patent are similar to ALL and 29 times higher than in ALL in Russia, 23 times higher in China, 7 times higher in the USA, and 3 times higher in Korea. The Russian and Chinese R&D in RES+ are particularly ineffective compared to the EU and Korea when measured by patents. The RES+ innovation performance of the EU is about 60 times higher than in the USA where more startups use patents but fewer survive the competition. These results show that the EU performs very well by innovations while patents are not essential features for the survival of startups. Although the EU policy is often perceived as bureaucratic, its innovation capabilities are higher than in other countries, aside Korea; note that the scale of innovations in Korea is a fraction of the EU. Moreover, the lower failure rates of startups in the EU imply lower societal costs of wasted R&D and business efforts. These results indicate that the idea of 'innovation systems' is not essential for innovations because the decision-makers about R&D subsidies cannot assess successful innovators when they decide about the allocation of budgets whereas patents are hardly relevant for the survival of the innovative startups. Moderately competitive business environments generate innovations while highly competitive environments are detrimental for the survival of start-ups.

An issue is what policies can invoke successful innovations in renewable energy. Given high innovation performance of the EU as a whole, the EU countries' innovation performances is assessed. The innovation performances in ALL sectors are estimated for 27 EU countries, excluding Croatia, and in RES+ sectors for 22 EU countries excluding Croatia, Cyprus, Lithuania, Latvia, Malta, and Luxembourg; those exclusions are caused by deficient data. The number of patents per research project (development performance) is multiplied by the number of survivors per firms' births (diffusion performance) to obtain the rate of the successful innovation process (innovation performance). The average R&D project is assumed to cost USD_{2010} 0.5 million, which is a typical cost of an engineering Ph.D. in the EU, including related equipment and staff; inter alia, it is twice higher than the average approved R&D proposal in the Horizon 2020. Appendix 3 shows the basic data: number of R&D projects, patents per project, survivors per birth, and the chances for innovations for ALL and RES+ sectors in each EU country.

Huge differences in the development and diffusion performances are observed across the EU countries. In ALL sectors, 8% of R&D projects delivered a patent in the EU, which varies from 2% in Portugal and Greece to 47% in the United Kingdom. Further, 10% of start-ups survive, which varies from −26% in Cyprus to +44% in Belgium. Minus means lower births than deaths as observed in 8 of the 22 EU countries, which indicate high-risk environments for the startups, contrary to Ireland, Belgium, and France where many start-ups survive. Given that these development and diffusion performances are not correlated across countries ($R^2 = 0.001$), the R&D and patents are not essential for the success of start-ups. Moreover the chance of patents across the EU countries is higher than in the Horizon2020 with laborious selection of R&D proposals which confirms deficient experts' assessments and a huge variation in the innovation performances across

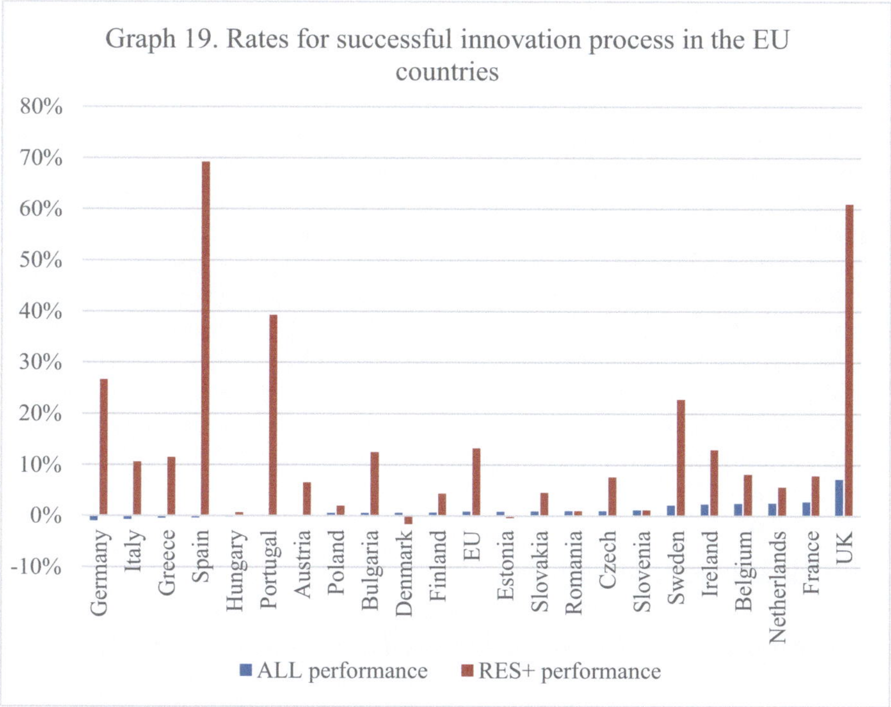

Fig. 4.4 Rates for successful innovation process in the EU countries

countries underpins that the experts' assessments of promising R&D for innovations is 'hot air'.

The innovation performances are strikingly higher in the RES+ sector with a huge variation across countries. About 18% of the R&D projects delivered a patent in the EU, this is only 3% in Estonia and 100% in the United Kingdom. While 74% of all start-ups survived in the EU, more firms failed than started in Estonia (−13%); more than 100% startups are statistically observed in Spain, Sweden, Czech, and Italy, which can indicate imports of firms or deficient data in those countries. The development and diffusion performances in RES+ are also not correlated across countries ($R^2 = -0.05$), which confirms that the R&D and patents are not essential for the innovation processes and that the experts' assessments of the R&D proposals in RES+ can be missed without causing problems to the funders and innovators. High innovation performances in RES+ are associated with the EU countries that pursue far-reaching policies in renewable energy, such as the United Kingdom, Spain, Portugal, Sweden, and Germany. Lagging policies are major impediments for innovation performances as the case in Netherlands and Poland. A low rate is found in Denmark with advanced policies but few survivors, which indicates that the far-reaching policy is not the only factor that generates high innovation performance because business environment, co-funding, and other factors are relevant.

Based on the data in Appendix 3, Fig. 4.4 presents the innovation performances listed in the ascending performances in ALL sectors.

The rates of successful innovation processes in ALL sectors are typically −1% to +1%; these are found in 22 of the 27 EU countries. While the R&D projects in Germany, Italy, and Cyprus have negligible rates of successful innovation, the rates are higher in the United Kingdom, Netherlands and France. The rates of success in the RES+ innovations are many times higher, typically 5–15%, with the spread from −2% in Denmark to 69% in Spain. Higher success rates of the innovation processes in RES+ than ALL sectors can largely be explained by the demand factors, in particular the demanding and supportive policies, and societal initiatives. Such demand factors determine high success rate of innovations in renewable energy.

The concept of an innovation system with subsidies for R&D and regulations for patents is hardly relevant for the survival of innovative startups because assessments for the subsidies fail and patents are not needed, while moderate competition and demand factors are decisive. Innovators in renewable energy are fostered when the far-reaching demands are posed by policies while tuning R&D to the expected societal demands is a key factor to success.

4.5 Financing Innovations

While innovators are usually convinced about their success due to better quality and price than competitors, the realisation and sales of inventions depend on the financing of innovations. The question is what options for the financing are available, given the uncertainties about their success as perceived by financiers. In conventional narrative, the innovative startups need external financiers because their own capital, including the ownership of patents, constitutes only a small asset compared to large expenditures in the innovation processes. The assumption is that the financial resources are scarce; These resources are either loans that must be repaid or equity that involves participation in profits. The innovators need collaterals as a guarantee for repayment if they aim to obtain a loan for the expenditures during the innovation processes. Alternatively, they need to attract participants that invest their capital in a new venture and share in profits or obtain public support through subsidies. As the innovator's collateral is usually small the loans is risky, which implies that innovators largely depend on the investors and public support. The innovators can obtain equity from risk-taking investors, called 'venture capitalists' that participate in the start-ups if innovation deems profitable within a few years after their investment. They can also involve public capital as a subsidy and participation. Figure 4.5 represents a profit function of the innovation process over time, with financers and typical rates of return on investment after correction for inflation, i.e., real returns.

The costs of R&D with patents, the proof of concept in a demonstration project, and pilot projects for the production growth of a startup enlarge over time and must be covered, whilst the risks decline because incomes in the future are better assessable. Though the time-to-market from the R&D to the production growth varies, it is rarely shorter than 5 years as R&D is laborious. Longer than 15 years is unacceptable

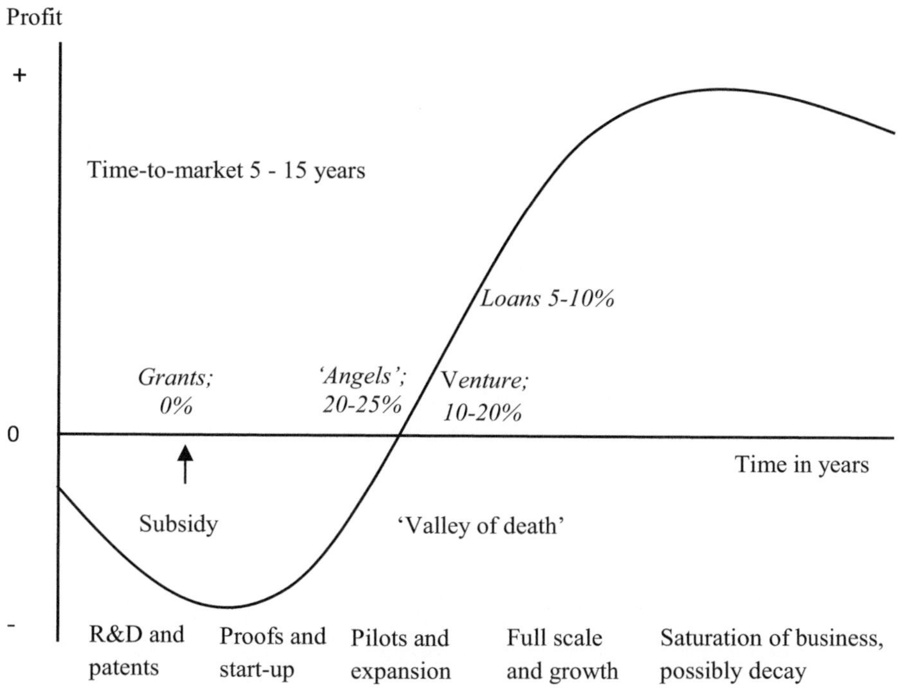

Fig. 4.5 Options for financing of an innovation process

to investors because their investments must be profitable whilst the present value of future sales declines; for example, the present value of income after 5 years at the usual 20% real interest rate is reduced to 40% of the original income. During that period, when costs are made but little income is generated called the 'valley of death', most innovators fail because they cannot bear all, increasing costs, whilst the income perspective declines over time because competitors can enter in markets and financiers become impatient.

However, this conventional narrative needs corrections. The venture capitalists are minor investors in innovators, besides a few successful ones that catch the eye. They rarely invest in the early phases of processes. Even in the pilot phase, they rarely approve of more than 3% of the submitted proposals for growth. In the EU, banks, innovators' funds, and governments were the main financial resources in all countries in 2017. Meanwhile, venture capital covered only 1–3% of all financing across the member countries. Even for the startups defined as 'young high-growth firms', the venture capital did not exceed 10%, except in Denmark where these percentages were higher (Invest Europe, 2019). Meanwhile, venture capitalists gain tax reliefs for investments in many EU countries which are assessed effective on the dubious grounds because without data on the tax expenditures and results for innovations compared to other investments (PWC, 2017). Regarding the negligible contribution of the venture capital to high innovation performance in the EU, these subsidies for the venture capital are considered wasteful expenditures (Coolman, 2021).

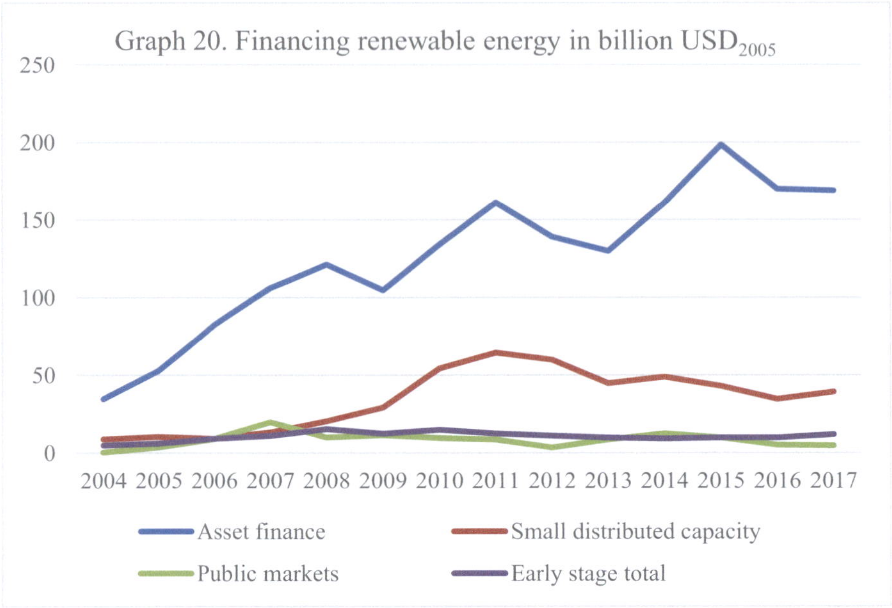

Fig. 4.6 Financing renewable energy in billion USD_{2005}

It is not different in renewable energy. Global financing of renewable energy is compiled in the UNEP/BNEF database (UNEP/BNEF, 2018). It is divided into the venture capital & private equity, government R&D, and corporate R&D which often finance startups, as well as the asset financing, small distributed capacities, and public markets finance which are mainly purposed for full-scale production. This division is somewhat arbitrary; for example, the International Renewable Energy Agency (IRENA) also shows the financial data in renewable energy (IRENA, 2020). Its division of financiers differs from the UNEP/BNEF database and estimate is nearly 50% higher with larger annual fluctuations than in the UNEP/BNEF data; however, analysis of the differences is beyond scope. Herewith, more cautious data of the UNEP/BNEF are used for the assessment of financing in renewable energy from 2004 to 2017 as shown in Fig. 4.6.

On average, about USD_{2005} 179 billion a year is spent globally on renewable energy during that period. Most expenditures covered low-risk activities. About 71% of all are assets of the vested firms which are usually based on loans delivered by banks and funds, 18% were small distributed capacities of trusts, individuals, and social organisations, and only 5% were public markets of stockholders because companies in renewable energy are rarely large corporations. The early-stage financing was only USD_{2005} 10 billion, which means that the investments in innovation are below 6% of the total because not all early-stage activities are innovative. These expenditures in innovations hardly changed in total during those 14 years, while their share in the total declined from 8% until 2010, to 5% in 2017. This indicates a declining interest of the investors in the innovators in renewable energy even though

R&D expenditures and number of patents in renewable energy increased in total and as a share of all energy between 2004 and 2017. A closer look into the early-stage investments reveals that the venture capital & private equity grew fast from less than USD_{2005} 1 billion in 2004 to more than USD_{2005} 9 billion in 2008, during the boom of R&D and patents in renewable energy followed by a fast decline to one-third 10 years later, during the decline of R&D and stabilisation of patents whilst the government and corporate expenditures on R&D grew steadily. So, venture capital is not a blessing for innovations in renewable energy but another way around: a boom of innovators provides opportunities for rewarding investments to the financiers.

The government is the main financier of innovations in renewable energy. It covers a part of the R&D expenditures and supports businesses with more than a hundred types of grants, credits, tax deductions, guarantees, and other financial policy instruments, which are often obtained in addition to another by firms (Ermen, 2007). Most instruments support the innovation processes before the start, while only a few after delivery of innovations. For convenience, the former is called 'subsidy' and the latter is 'guaranty'; they are shown in Fig. 4.5 by the upward shift of the profit line for subsidies and the inward shift for guarantees. While subsidies reduce the innovators' costs, they shift the burden of risk from private investors to the public without sharing in eventual profits. In order to reduce the risks for public investments, the subsidisation involves detailed assessments of the innovators' promises by experts. The guarantees for sales prices, purchases, or capital expenditures reduce the financiers' risks because they create more stable markets, without direct public investments. Their purpose is to attract capital and reduce the interest rates of loans and participations without detailed assessments. Note that lower interest rates improve the present value of income in a logarithmic manner which implies that even small changes in the interest rates have large impacts on innovation processes; explanation of the present value can be found in the Appendix 1.

Guarantees in renewable energy were introduced in the Germany policy in the early 2000s, called 'feed-tariffs'. This regulation obliged operators of networks to purchase renewable energy at the prices set by the regulator well above the market price of lower cost fossil fuel. These feed-in tariffs encourage supplies of renewable energy without specifying producers or consumers, which creates a large, market for renewable energy and reduces price fluctuations, thereby the investors risks. Meanwhile, the regulators are not constrained by the budget and fiscal limits because all financial transfers are market-based (Böhme & Dürrschmidt, 2007). Within a few years, this regulation disseminated across the EU with different tariffs for the renewable energy resources, years of guarantee, the scale of supply, and other conditions (Klein et al., 2008). The feed-in tariffs in the EU were effective, when measured by the number of firms and employment per euro spend on these guarantees (Krozer, 2017). Meanwhile, feed-in tariffs refer to the guarantees and subsidies used interchangeable when purposed to boost renewable energy. Note that guarantees were used for long in the procurement and recycling of old paper, textiles, and construction materials (Edler & Georghiu, 2007), take-back of discharged electronic and mechanical products (Iszak & Edler, 2011), as well as in agriculture, public services, and other domains.

Another source of capital refers to the entrepreneurial capabilities in generating financial resources through crowdsourcing and crowdfunding. The crowdsourcing means involvement of users in innovation processes, also called the 'user innovations' or consumer innovations. Measured by time spent by consumers on innovations in the United Kingdom alone, the crowdsourcing valued at €3.8 billion in 2012 (Hippel et al., 2012) was larger than the European venture capital of €3.5 billion in 2013 (Statista, 2021). Examples of such consumers' involvements in solar cars and solar boats show fast effect-increasing technical change measured by the speed of cars and boats with a large business spin-off. Small grants for entrepreneurs with crowdsourcing without any expert assessments proved to be more effective by the number of novelties, and less costly per novelty compared to the EU subsidies with the expert assessments (Krozer, 2015, Chapter 5). Similar results are observed with the use of innovation vouchers, which are small grants from USD 4000 to USD 25,000 for the entrepreneurs dedicated to involving researchers in their innovation processes, thereby linking the institutional R&D to entrepreneurs (Spiesberger & Schönbeck, 2019). Crowdfunding refers to the consumers' participation. These participations nearly doubled every year from USD 11 billion in 2013 to more than 418 billion in 2017 (Ziegler et al., 2020). Whilst some countries as Germany, Italy, and Spain in the EU maintained a long-standing tradition of communal and cooperative energy enterprises, the citizens' participation and cooperatives emerged in other countries after the privatisation of energy production; for example, in Belgium and Netherlands. These financing tools enrich the conventional models of the loans and shareholding with the stakeholders' participations in ventures (Tyl & Lizarralde, 2017).

Possibilities for generating risk-taking participations in renewable energy grow when innovators consider a broad scope of the policy instruments and the stakeholders' participation, while conventional venture capital is scarce and costly.

4.6 Conclusions

Possibilities for generating and steering energy technologies are assessed based on the mainstream assumption that R&D delivers inventions protected by patents, which enable the start-up of firms that pursue profits in market competition, but in each phase, failures are experienced. A concern is whether novel energy technologies are plausible within a few decades, because the duration from inventions to innovations, and diffusion can be longer. Therefore, the lead-times of 29 basic energy technologies for conversions of biomass, coal, chemicals, oil, gas, hydro, nuclear, geothermal, wind, and solar energy are estimated. The typical lead-times of innovation processes span six to ten decades, which is seen in 19 cases of basic energy technologies. Only in four innovation processes, the lead-times were shorter than six decades, all of them were driven by military interests, in six cases, the lead-times were longer than ten decades, mainly caused by societal resistance against explosions, pollution, and suchlike hazards. Regarding those lead-times, breakthroughs should not be expected within a few decades. Moreover, precaution regarding the harms of technologies can avert time-lags that cause even longer lead-times of innovation processes or even stagnation of the innovation diffusion because an increasing societal risk-aversion can be expected as welfare grows.

Looking at a shorter timeframe, the R&D expenditures in renewable energy and their results in patents generated a larger business interest than in all energy R&D and patents and all sectors. While there is not much business interest for the energy R&D and patents, measured by their shares in all R&D and patents across countries during 2008–2017, the renewable energy R&D and patents attract high business interest in particular when combined with storage and ICT. However, this business interest is sensitive to energy prices as high fuel prices enhance that R&D and patents. That R&D and patents are concentrated in the USA, Japan, and EU, whilst mid-income and low-income countries catch up and R&D in a few of these countries efficiently generate patents in renewable energy. It indicates that the accessibility of renewable energy increases across countries.

High business interest does not guarantee profitable innovations because the results of R&D measured by patents also depend on the possibilities of the startups to survive in the competition. Experiences in the EU show that the expert assessments of excellent R&D proposals are 'shots in the dark', measured by patents. While China, Japan, Korea, Russia, and the USA perform better than the EU by this indicator, the EU performs better measured by startups per patent and survivors per startup, the EU performance is nearly as good as smaller Korea. The main advantage in the EU for startups is the business environment of moderate competition compared to other countries. Within the EU, the chance that a large R&D project generates a profitable innovator is close to nil across the countries. So, the ideas that the expert assessments can allocate subsidies effectively and that the patents enable the startups to compete have little foundation in practices. It is not better in the other countries, given the high performance of the EU. The chances that an R&D project generates a successful startup increase by 10 to 15 times across these EU countries when policies pose far-reaching demands for renewable energy as observed in a few EU countries. The rate of success increases because opportunities for profitable innovations are generated. The demands foster successful innovators; the expert assessments for subsidies and patent regulations rarely do.

Given the low chances of successful startups, financing is a major challenge. A popular narrative is that the risk-taking venture capitalists invest in promising startups; this is used for the justification of tax exemptions that are obtained by these investors. However, most startups are financed by banks, starters own funds and policies; they are also less sensitive to market fluctuations than the venture capital. Various policies for financial support of innovative renewable energy can be divided into subsidies delivered before innovations or guarantees provided after the innovation. The subsidies reduce the costs of innovators but shift the risks from private investors to the public without participation in profits and knowing that the chance to subsidise success is nearly nil. Guarantees reduce the public risk and provide market stability for the startups, which have been successful for the innovators in renewable energy in several EU countries as feed-in tariffs. Modern financing methods, based on the idea that the entrepreneurial capabilities are decisive for successful innovations, refer to crowdsourcing and crowdfunding which are ventures based on the participation of citizens. Meanwhile, these modern financing methods are much larger by the financial volume than the venture capital and can be facilitated by regulations for large-scale investments in renewable energy.

The innovation processes in new energy technologies encompass several decades, while innovations in renewable energy can be enhanced by far-reaching policy demands in combination with price guarantees and modern financing methods based on the citizens' participation.

Chapter 5
Innovating in Renewable Energy

Why has modern renewable energy grown fast during the last fifty years, although it is considered high cost, low energy density, and intermittent? An answer refers to the decision-making of individuals, organisations, and authorities. Data on investments, consumption, prices, subsidies, social initiatives, and firms in the past were used to assess the impacts of energy prices, policy support for fossil fuels and for renewable energy as well as entrepreneurial initiatives in modern renewable energy. A few patterns in the countries' decision-making about modern renewable energy are underpinned. The cost-reducing technical change as a result of large investments in wind and solar energy is estimated and the benefits of applications are assessed.

5.1 Introduction

When modern renewable energy emerged in the 1970s due to the path-breaking work of innovators in the USA, Japan, and the EU, not many expected fast-paced growth because these innovations were costly compared to energy generated with fossil fuels and their performance was deficient. Furthermore, policies were preoccupied with struggles for the cheapest fuel resources and expansion of nuclear power, whereas those innovators in geothermal heaters, windmills, solar cookers, solar panels, dung gas stoves, and suchlike products were usually dismissed as 'eco-freaks', not welcomed in the neo-Marxist, liberal and conservative political families at that time. Nevertheless, those innovations were turned into the commercial mainstream, contrary to nuclear power that was embraced by all those ideologies as the energy technology but stagnated after the 1980s. Why did modern renewable energy, based on geothermal, wind, and solar energy, and biofuels have grown during the second half of the 1900s, although these resources are intermittent, whereas traditional renewable energy and fossil fuels provide continuous, energy-dense resources? This focus is justified by the high growth rates of modern renewable energy in energy consumption throughout the late 1900s and the early 2000s, in

Y. Krozer, *Economics of Renewable Energy*,
https://doi.org/10.1007/978-3-030-90804-1_5

particular the fast growth of wind and solar energy towards the large-scale generation of electric power. Reasons for this turnaround are discussed with regard to the mainstream thinking about market prices, evolutionary focus on policies, and behavioural one on the entrepreneurial initiatives mainly using the statistical data starting from the early-1970s and ending in the late-2010s.

The mainstream argument is that the increase in the price of fossil fuels during the 'oil shock' of 1979–1985 and 'demand shock' 2005–2015 has triggered alternatives for fossil fuels in high-income countries. The subsequent expansion of renewable energy technologies reduced costs per energy unit due to the economy of scale that entailed price parity with the incumbent technologies. However, this reasoning does not explain why costly innovations were generated rather than cheaper hydropower and nuclear power. Furthermore, tar oil, shale oil, and gas fracking were viable options by the late-1900s, while their production grew only in North America. The complementary arguments are that the policy support has reduced the costs of renewable energy and that the military and telecommunication interests within the state administrations have enhanced renewable energy in remote areas; a nexus of interests in the development of energy technologies is observed during the late 1900s, but no references are given for the earlier period (Block & Keller, 2008). A shift in social values is also pinpointed. Attention to efficiency through the larger scale of technologies would turn toward so-called 'downscaling of technologies', which was popularized with the phrase 'small is beautiful' from the 1970s onwards (Schumacher, 1973). From this perspective, 'appropriate' technologies tuned to low-income countries and ecological interests were pursued and nuclear energy was scrutinised (Ulrich, 1979), whilst the perceptions about what is 'appropriate' diverged throughout fifty years from 1960s on into directions ranging from global geoengineering to distributed energy systems (Symons & Karlsson, 2015).

Herewith, those argumentations are discussed with use of the statistical data on energy consumption, and policy support, followed by assessments of the early social initiatives and how they turned them into startup firms in the EU, types of decision-making across countries, the impact of investments on the cost-reducing technical change and finally an assessment of the benefits.

5.2 Prices and Modern Renewable Energy

From the mainstream perspective, one would expect that high international prices of fossil fuels triggered a reduction of energy consumption along with substitutions of costly fossil fuels for lowest-cost resources, and invoked growth of cheapest renewable energy. However, this is not observed in volumes of fossil fuels and renewable energy from 1965 to 2019, based on the British Petroleum (BP) database. This time-span of 55 years encompasses the periods of low international prices of fossil fuels during the 1960s until 1973 and from 1985 to 2004, as well as high prices during the 'oil shock' from 1974 to 1985 and the 'demand shock' from 2005 to 2015. Figure 5.1 shows the prices of coal, oil, and gas, as well as the average prices, weighted for the

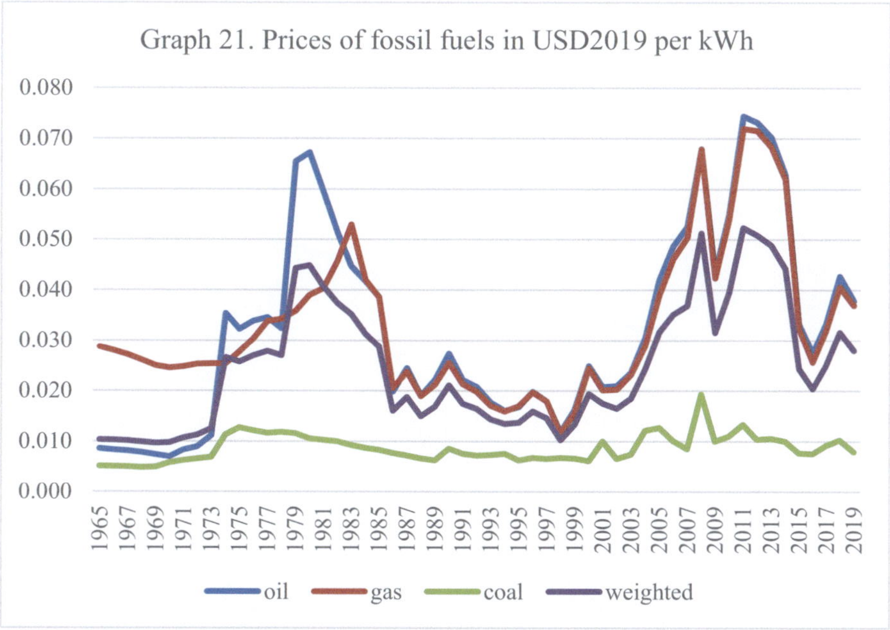

Fig.5.1 Prices of fossil fuels in USD2019 per kWh

consumed volumes of coal, oil, and gas. All prices are in dollar-cent in constant USD_{2019} per kWh using BP data.

Throughout the time span from 1965 to 2019, the coal prices fluctuated mildly between 0.6 dollar-cent and 1 dollar-cent per kWh whilst the combined oil and gas prices swung from less than 1.0 dollar-cent during the 1960s to nearly 7 dollars-cent in 1980, and even higher in the early 2010s. However, the global consumption of coal grew slower than oil, even slower than gas, contrary to higher prices per energy unit, which is underpinned with data below. Possibly, the energy services possess characteristics of the 'Veblen goods' meaning that higher prices go along with higher status, or because oil and gas are more convenient in uses. Nuclear energy hardly influenced international prices as it was of a smaller scale. The weighted, average prices of fossil fuels doubled from 2.5 dollars-cent per kWh in the 1960s to 5.0 dollars-cent per kWh during 1979–1984, followed by a decline. They doubled again to 5.5 dollars-cent per kWh during 2005–2015, followed by a decline to the 2004 price level. Meanwhile, the peak prices of oil and gas increased about fivefold. Moreover, the average prices per kWh increased along with larger energy consumption, which is not in line with the neoclassic argumentation, possibly because the income-elasticity is higher than price-elasticity of demand, meaning that people spend more on energy per unit of income to price because energy services are perceived as basic features in consumption.

Similarly, the decreasing CO_2 emissions can be expected based on higher CO_2 equivalent prices, measured by the CO_2 equivalents of energy resources. Using those energy prices per kWh, the CO_2 equivalent prices of energy resources are estimated. This is done with the BP energy prices in USD_{2019} per kWh, consumption per resource

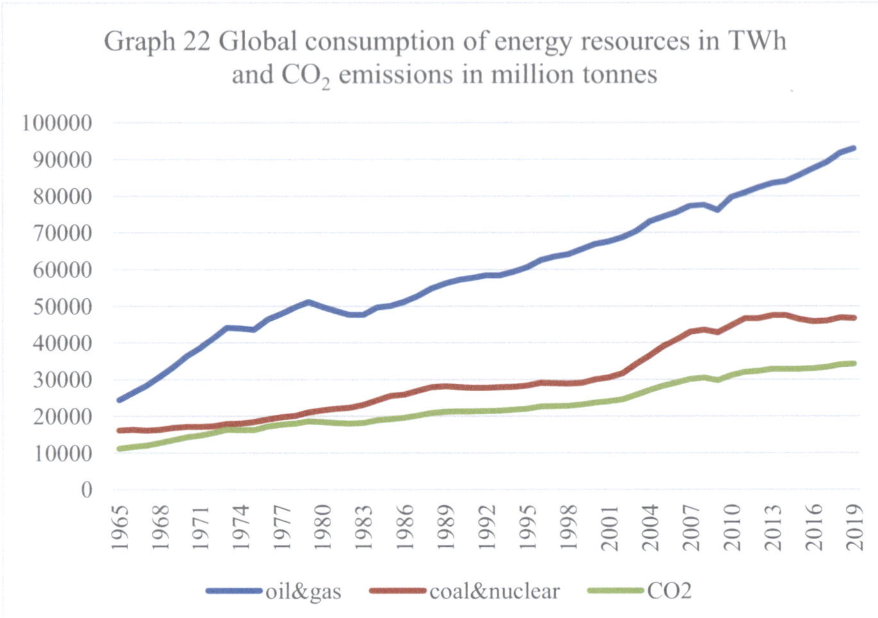

Fig. 5.2 Global consumption of energy resources in TWh and CO_2 emissions in million tonnes

in kWh, and the IEA coefficients of energy resources in kg CO_2 per kWh; the estimated CO_2 emissions with the IEA coefficient hardly deviates from the BP estimate. The weighted CO_2 equivalent prices were USD_{2019} 132 per tonne during the 'oil shock' 1974–1985 compared to USD_{2019} 40 per tonne during 1965–1973 and USD_{2019} 161 per tonne during the demand shock 2005–2015 compared to USD_{2019} 66 per tonne during 1986–2014; all these are annual averages during those periods.

The impacts of twice to threefold higher energy prices during those shocks on energy consumption and CO_2 emissions were low. Consumption of all fossil fuels and the emissions were steadily growing with minor downward ripples in the scale of consumption en emissions during high energy prices. Note that the economic projections of energy consumption and CO_2 emission usually assume that prices and taxes cause the declining consumption and emission (e.g., Mercure et al., 2014). Figure 5.2 shows the consumption of oil with gas and coal with nuclear energy in TWh, as well as CO_2 emissions in million tonnes from 1965 to 2019 based on the BP data.

While the annual average prices of fossil fuels in USD_{2019} increased from 1.1 dollar-cent per kWh during the period 1965–1973 to 2.7 dollars-cent throughout 1974–2019, global energy consumption grew steadily. Even increases in oil and gas prices during the price shocks in 1974 and 2005 caused only temporary dips in 1979 and 2008. Low sensitivity of energy consumption with respect to prices has also an impact on CO_2 emissions. For example, had the dip in 1979 continued after 1982, the global energy consumption would be 59 PWh (214 EJ) in 2019 instead of 162 PWh (584 EJ) and climate change would not be a major issue because CO_2 emissions would be nearly half lower of emissions throughout those 40 years. However, the growth of energy consumption was not sensitive to high resource prices, which casts doubts about effects of the mainstream proposal for carbon taxes on mitigation of climate change.

Furthermore, the substitution of oil and gas for cheaper coal and nuclear energy did not occur. The substitution could be expected based on the mainstream thinking because the coal prices were five times lower than the oil prices during the shocks and nearly twice lower during low oil prices. Despite lower coal prices during the price shocks, coal grew slower compared to oil and gas. Coal is increasingly used for the electricity production, which illustrates specialisations that generate complementarity rather than substitution in energy production and consumption. Moreover, the electrification grew fast because 3.0% annual average from 1985 to 2019 based on BP data which was largely based on coal and nuclear resources; by 2019 36% of the electricity generation was based on coal and 10% on nuclear resources. Note that the World Bank data from 1971 to 2014 shows half lower electricity production and only 2.3% annual average growth, along with a similar share of coal in energy production. Despite lower prices and the fast-growing electrification largely based on coal, the consumption of coal and nuclear power grew slower than oil and gas. As a result, oil and gas consumption increased from about 60% of all fossil fuels in 1965 to 76% in 2019, when cheaper coal covered 20% and even cheaper nuclear resources did not exceed 4% of all fossil fuels.

Sluggish energy-saving and substitutions between the fossil fuel resources are associated with growing CO_2 emission albeit slower growth than coal and nuclear energy. Changes in the average price of fossil fuels, weighted for consumption, are moderately correlated with the changes in the scale of CO_2 emission ($R^2 = 0.3$), but this price elasticity of CO_2 emissions is inconclusive because it varies per period. Higher prices had a small impact on the growth of CO_2 emission while the price-equivalents of CO_2 increased from the annual average of USD_{2019} 65 per tonne CO_2 during 1990–2004 when oil prices were low to USD_{2019} 165 per tonne CO_2 during 2005–2015 when oil prices were high. This CO_2 equivalent is based on the weighted average of CO_2 emission per kWh of oil, gas, and coal. The average price, weighted for the consumption of those fuels, increased nearly precisely USD_{2019} 100 per tonne CO2 during high prices from 2005 to 2015 compared to 1990–2004; the increase was about 50% higher during the price peak of oil in 2011. However, the growth of CO_2 emission hardly declined during high fuel prices. Possibly, subsidies for energy consumption dim the consumer prices and the consumption volume because consumer habits and technologies are persistent. Also, the decreasing share of energy costs in income can temper the price incentives. Whatever the reason, the mainstream assumption that higher prices of fossil fuels substantially reduce CO_2 emission is not valid. As the prices mechanism is not as forceful as it is often expected, excessive expectations in policies about the positive effects of the CO_2 pricing on emission reduction should be tempered, unless combined with other policy instruments.

Whether high oil and gas prices generate renewable energy is also disputable. High prices of the 'oil shock' in the mid-1970s did invoke renewable energy, but not specifically the cheapest traditional resources based on hydropower, biomass, and waste. Hydropower grew faster than fossil fuels after the oil shock. This growth was hampered when the fuel prices declined in the late 1980s. Conversions of biomass into alcohols grew in a few countries as a few oil companies invested in joint ventures for biomass processing, but the companies divested after the decline of the oil prices in the mid-1980s. For example, Shell was one of the largest investors in wood processing and forecasted 2.5% annual average growth of biomass 1980 to 2000, whilst the UN forecasted 3.5% – 4.5% if the cost of bio-alcohols attains USD 50 per b.o.e. while bioenergy grew even faster during the 1980s despite the divestments of oil companies (Krozer, 1987).

Costlier modern renewable energy emerged based on the geothermal, wind, solar, and biofuel resources in the ascending order of their costs per energy unit and historical order of introduction in markets. That ranking of unit costs is similar in authoritative databases and the ranges and averages of the unit costs differ (EIA, 2020; Lazard, 2020; NREL, 2021b). While the oil and gas prices swung from 1980 to 2019, hydropower and biomass grew by 2–3% annual average, which was similar to the growth rates of fossil fuels. Meanwhile, geothermal energy increased by 7%, wind energy by 34%, solar energy by 32%, biofuels by 43% annual average, all based on the BP data. High growth rates of wind and solar energy were attained despite their intermittent property, whereas hydropower and biomass, as well as fossil fuels, could deliver continuous supplies. Moreover, wind and solar energy rivalled with coal in the generation of electricity whilst the coal prices hardly increased. Vice versa, biofuels remained small scale, though they were rivals to high oil prices. The price impact on the production growth of modern renewable energy is not significant statistically as only the low growth rates of geothermal energy and the price changes of weighted fossil fuels are correlated, the growth of wind and solar power and these price changes are not correlated. Furthermore, the international prices of fuels should have had a global impact; however, the impacts differed across countries. For example, solar energy in the Middle East with the highest solar irradiation in the world and large financial capacities remains small scale, or larger wind resources in thinly populated Russia are hardly used compared to the intensive use in the densely populated EU.

Possibly, policies are relevant for the growth of modern renewable energy. In evolutionary thinking, the policy support would generate large-scale production of modern renewable energy entailing cost-reducing technical change. However, the traditional renewable energy remained larger scale than wind and solar energy – 30 times larger in 2019 – but the modern one grew much faster. Based on behavioural thinking, it could be argued that the societal initiatives generated beneficial qualities for stakeholders. The growth of modern renewable energy would be driven by societal initiatives if they could convince the decision-makers about the value-added energy services with modern renewable energy; for example, generation of electricity by communities entailing startups and jobs. Whether those hypotheses can explain the faster growth of modern renewable energy compared to the other energy resources is assessed in the subsequent sections.

High international prices of oil and gas hardly increased energy efficiency, did not trigger the substitution of oil and gas for cheaper coal and nuclear energy, neither much energy saving, nor much CO_2 emission reduction. They did invoke modern renewable energy though the prices were not the only drivers of its fast growth from the 1980s onward.

5.3 Energy Subsidies

In the evolutionary argumentation, policies determine energy production and consumption directly through investments and operations of the public firms as well as through regulations. No doubt that policies influence energy production and consumption in general. A question is whether the energy policies have encouraged the growth of modern renewable energy, or they impeded it when aimed to support their domestic energy production and consumption. In many countries, domestic energy production is wholly or partially owned by the states. Although in the last decades of the 1900s, many countries privatised production, the states kept infrastructure in their own hands. Domestic energy consumption is private when consumed at households and in the private organisation whereas policies influence that through policy instruments, when the authorities aim to attain various, competing policy goals; for example, accessible energy, and low pollution.

Within that institutional context, this assessment focuses on the financial instruments, particularly on the financial support of private interests. This support encompasses the governmental expenditures shown on the state budgets, as well as larger support with credits, tax exemptions, soft loans, and other administrative interventions not shown on budgets, which are usually called 'on-budget' (or 'pre-tax') 'off-budget' (or 'post-tax'), respectively. 'On budget' support can be found in the statutory reports of the national, regional, and local authorities because must be approved by the institutions that control the government, whereas the off-budget support, which is woven into many sectoral accounts, is laborious to assess. For the sake of convenience, all kinds of financial policy support are called subsidies. If energy production or consumption are subsidised, that consumption is enhanced because the costs decrease. So, if fossil fuels are subsidized, they become cheaper relative to renewable energy, which implies that renewable energy technologies are obstructed and CO_2 emission is enhanced compared to nil subsidies.

Attempts are made to estimate all energy subsidies. A comprehensive global estimate is done by the International Monetary Fund (IMF) with the so-called 'price gap' method. This gap covers the difference between the consumers' expenditures and producers' costs for the estimate of consumer subsidies, plus the difference between the producer expenditures and costs of all resources for the producer subsidies; note that it is a laborious method with many uncertainties but the best we can get at this moment. The IMF estimated global subsidies amounting to USD 1900 billion in 2012 (Clements et al., 2013). This estimate implies that subsidies cover about 29% of the global energy market. Subsequent IMF estimates underpinned the fact that most subsidies supported fossil fuels. When the damages caused by CO_2 emissions were included by the IMF, the estimated subsidies of fossil fuels were nearly three times higher (Coady et al., 2015). These authoritative assessments show that production and consumption of fossil fuels are heavily supported.

Large subsidies are off-budget, mainly taxes. They are usually distributed in favour of large-scale producers and consumers because the energy taxation in most countries is regressive, meaning that larger energy consumption is less taxed per energy unit. A global review of the tax exemptions is not found. In the EU, as an example of regressive taxation, the tax exemptions for large-scale energy consumers compared to small-scale ones were about € 118 (USD 149) billion annual average during high fuel prices in 2005–2015, which is more than 18% of all energy costs. Each country has different taxations. Nevertheless, the taxes on household consumers per unit of energy are usually nearly ten times higher than the energy taxes on medium-scale business consumers, and up to thousands of times higher than the energy taxes on the large energy businesses and the governmental institutions (Krozer, 2015, Chapter 10). A more recent estimate for 2019 shows larger, € 137 (USD 174) billion energy subsidies (Ferguson, 2021). Whether these subsidies have increased, or that difference is caused by different assessment methods cannot be verified, which is important because the official EU policy aims to eliminate energy subsidies, in particular all subsidies for fossil fuels. Besides the tax exemptions, there are policy agreements with large, energy-intensive firms about discounts for the network cost and support of energy consumption. In a result, some energy-intensive large firms hardly pay for their energy consumption. For example, the most energy-intensive refineries, chemical, and metal industries in the Netherlands pay three times lower energy prices than similar industries in Belgium, and even up to 10 times lower prices than in Germany (Velthuijsen, 2018). Negligible energy taxes on consumption in the energy-intensive industries in the Netherlands are paid by the highest taxes on the households energy consumption in Europe (European Commission, 2020).

On-budget subsidies are also substantial. The IEA-OECD shows that on-budget global energy subsidies steadily increased from USD 342 billion in 2007 to USD 544 billion in 2012, followed by a decline to USD 262 billion in 2016, but these data cover a limited number of countries; for example, the USA and EU are not covered although they participate in the IEA and OECD. The IRENA estimates show larger global subsidies, on budget in 2016. Its estimate is USD 447 billion for coal, oil, and gas, half of it for oil, as well as USD 169 billion for nuclear energy, USD 128 billion for renewable energy, and USD 38 billion for biofuels in transport (Taylor, 2020). This IRENA estimate of USD 782 billion on-budget subsidies, 79% of that for fossil fuels, is nearly three times higher than the IEA-OECD estimate. Nevertheless, on-budget subsidies were presumably larger because solely the USA and EU subsidies for energy exceeded USD 460 billion in 2016 based on the official data. The EU data indicated that about 50% of its USD 187 billion (€ 169) on-budget subsidies in 2016 were for renewable energy (EC, 2019). A difficulty with such assessments is that all countries are reluctant to provide reliable information about subsidies because the policy support of domestic producers and consumers contradicts with the international agreements in the World Trade Organisation about the erasing barriers for trade caused by protectionist actions; therefore, large subsidies can trigger sanctions that cause social costs which countries tend to avoid through obscure data and deficient verification.

Contrary to high taxes on household energy consumption in most high-income countries, household purchases of commercial energy resources are subsidised in many mid- and low-income countries. As the commercial energy resources are usually fossil fuels in those countries; as an effect, the consumption of fossil fuels is subsidised. High subsidies for the household consumption of fossil fuels are usually based on the argument that they foster energy access because people with low income can benefit from cheaper energy. A question is whether this argument is valid, or it is an occasional argument; for example, during elections. If this argument is valid, the countries' subsidies would increase when the prices of fossil fuels increased in order to compensate for higher costs, and the subsidies would decrease when these prices decreased because less necessary for lower incomes. However, below is underpinned that the subsidies for household consumption of fossil fuels have little to do with the compensation for high prices.

The impact of energy subsidies is assessed with respect to the fuel prices using the IEA-OECD data on energy subsidies for oil, gas, coal, electricity, and transport fuels in 41 countries, whereas the subsidies in transport are combined with oil subsidies and electricity is not considered because it involves mainly conversions of those fossil fuels (IEA subsidies, 2021). Within the decade of 2010–2019, the periods of high prices 2010–2015 and low prices 2016–2019 are specified; note that the average prices of oil and gas during the latter period decreased to one-third of the entire period 2010–2019. The subsidies data are matched with BP production data on coal, oil, and gas in 32 countries because some production data do not match with the subsidy data. Each combination of a country with an energy resource is considered as a case.

It is found that the energy subsidies increased in 35 cases during the whole period, and in 27 cases during the decreasing prices. Meanwhile, the subsidies decreased during the whole period in 22 cases, and in 4 cases during the decreasing prices. Therefore, in 31 of the 57 cases, the subsidies increased during declining market prices which indicates that higher subsidies for fossil fuels were not primarily purposed to compensate high fuel prices. Moreover, high correlations ($R^2 > 0.8$) between the growth of fossil fuels consumption and the increasing subsidies are found only in three cases (China and Indonesia for gas, and Malaysia for oil), and in four cases of the decreasing subsidies (Ecuador, Kazakhstan, Thailand, and Ukraine for gas). Only the former three case indicate and impact of subsidies for fossil fuels on household consumption, but in 87% of all cases there is low or nil correlation between the subsides and consumption of fossil fuels. Furthermore, the subsidisation rates of energy also vary from 0% in Taiwan, to 91% in Venezuela – it is nil to nearly all the cost of energy – as they are unrelated to incomes per capita. In effect, the subsidies in the income per capita vary from USD 1.1 in Ghana and Vietnam to USD 926 in Turkmenistan where the income per capita is a few times higher. Generally, these subsidies are ineffective, and the budgets can better be allocated in support of lower incomes through jobs, basic income, tools for energy-saving and renewable energy, and other income-generating activities.

Given that most subsidies support fossil fuels, the subsidisation of energy can impede renewable energy. Using IEA-OECD data, it is estimated if the scale of subsidies for fossil fuels has impacts on renewable energy. Therefore, the energy subsidies per GDP in 2016 are correlated with the share of renewable energy in energy consumption, based on World Bank data. For those 40 countries – excluding Taiwan as it is absent from the GDP data of the World Bank – a moderate, negative cross-countries correlation is found ($R^2 = -0.5$), which implies that these subsidies often impede renewable energy. It is also estimated that a one percent subsidy in the GDP reduces more than one percent of renewable energy in energy consumption in most countries. The impacts below unity are found for the largest oil-producing countries where the consumption of renewable energy is low. Possibly, most oil-producing countries can afford energy subsidies and small-scale renewable energy. For example, in Venezuela, one percent subsidies in GDP increased more than one percent renewable energy consumption as the scale of this consumption is negligible.

While that policy support of fossil fuels enhances CO_2 emission, an instrument aiming at the emission reduction is the CO_2 tax because a price on CO_2 is assumed to invoke energy-saving and generate a price advantage for renewable energy compared to fossil fuels. In the mainstream theory, a tax per emission unit set at the marginal damage cost delivers optimal pollution reduction; it is when neither polluters nor victims are worse-off. As the CO_2 damage costs vary from USD 5 to 395 per tonne in more than a hundred reviewed publications, whose mean is USD 104 per tonne (Tol, 2005), it is disputable what tax generates an optimum outcome. For conveniences, tax at the mean level of USD 104 per tonne CO_2, can be considered close to the optimum; it can be noted that the multiplication of the global CO_2 emissions and this mean show the similar result to the estimate of global damages by the IMF mentioned above. Regarding these damage costs, the IMF recommends the CO_2 taxes of USD_{2018} 30 per ton CO_2 for mid and low-income countries and USD_{2018} 70 per tonne CO_2 for high-income ones onwards (Lagarde & Gaspar, 2019). Note that the proposed tax of USD_{2018} 70 per ton CO_2 is equivalent of an additional USD_{2018} 24 per b.o.e., based on 2.53 kg CO_2 per kg oil, which is half lower than the increase of the oil prices during the periods of high oil prices.

Those taxes have not been reached. The number of CO_2 taxes increased from the 1990s on, but nearly all were below USD 10 per tonne CO_2 and only 13% of CO_2 emissions across countries were taxed in 2015, of these nearly one-third of all taxes were regional and local (World Bank, 2016). In 2015, the CO_2 tax in Sweden was about twenty dollars higher than that mean damage cost, similar to the price increases of oil during high oil prices, in Switzerland similar to the mean and in Finland similar to the IMF recommendation; all other taxes were substantially lower (Keen et al., 2019). So far, most countries did not introduce any taxes and litigation of climate change is in infancy (Solana, 2020). If all global CO_2 taxes are divided by all CO_2 emissions in 2019, the average tax is USD 2 per tonne (O'Mahony, 2020). Therefore, the taxation of CO_2 emissions is negligible whilst the subsidisation of CO_2 emissions through on-budget and off-budget support of fossil fuels is substantial; this is contradictory to the international agreements about mitigation of climate change.

Whether high CO_2 taxes enhance renewable energy and reduce CO2 emissions is uncertain because Sweden, Switzerland and Finland are not among the largest producers of modern renewable energy, though Sweden is a large producer of bioethanol. Furthermore, experiences during high international prices of fossil fuels showed that these prices had a limited impact on the CO_2 growth. Given that the recommended CO_2 taxes are below those high prices, some impacts can be expected only if such taxation is combined with the elimination of subsidies for fossil fuels with compensation for low-income; for example, support of the renewable energy consumption. Whether the impacts of market prices on CO_2 emission reduction differ from the impacts of CO_2 prices set by taxation is not assessed and reviews of experiences are not found. It is plausible that the impact of the policy-based prices are substantial compared to the market prices because the policies also send a normative message which influence expectations about the trend for the future.

Another policy instrument is the CO_2 pricing with emission cap and trading of emission allowances distributed across sectors, called the emission trading system (ETS). In theory, it can limit emissions of the growing economies through the cap and generate an efficient emission reduction through the market transactions. It is assumed that in the growing economies a cap creates scarcity for the CO_2 allowances followed by an increase of the CO_2 prices. Inspired by the NOx and SO_2 emission trading in the USA and Kyoto agreement, the EU introduced the ETS for CO_2 in the early 2000s. From 2008 onwards, the EU policy distributes yearly certified allowances for CO_2 emissions among large firms, while firms were allowed to trade between those firms with a surplus of allowances and ones without sufficient allowances to cover their emissions. Insufficient allowances relative to emission were penalised. However, many allowances were granted based on emissions of the past rather than based on the 'polluter pays principle', which is the cornerstone in policies, and these allowances were granted for free to the largest polluters. In effect, those firms that reduced emissions in the past were penalised, and a stock of unused allowances could be sold at a price without any costs, which generated 'windfall profits' by the largest polluters considered unfair. This, so-called 'grandfathering', obstructed emission reduction because too many allowances are introduced compared to the actual emissions and granting of the allowances for free undermined the CO_2 prices.

Despite these deficiencies, the ETS has contributed to the CO_2 emission reduction, and when the EU Commission intervened in the emission trading by taking away a part of allowances, the allowance prices are restored. Regarding positive results, the ETS system is broadened in 2021 toward nearly all businesses and transport. Contrary to the grandfathering, there have been calls for more public interventions by reducing the allowances in line with the EU targets for 55% emission reduction in 2030 and elimination of CO_2 emissions in 2050. Furthermore, a possibility that citizens can purchase allowances is advocated which would enable to increase the CO_2 prices through civil actions. So far, the decision about the grandfathering, or another distribution method of the allowances, and about the policy interventions in the ETS market is pending, but the possibility of purchases by citizens is not included in the EU policy.

Globally, by 2020, that CO_2 emission trading covered about 9 billion tonnes, of which 8 billion tonnes was in the EU and 1 billion tonnes divided across Korea Republic, Switzerland, and several regions in Canada and the USA. The transaction prices fluctuated from USD 5 to USD 30 per tonne (REFINITIV, 2021). Those prices fluctuated strongly a few years after the introduction of the ETS in the EU, followed by convergence across those countries and regions toward USD 30–40 per tonne; which approximates the unit costs of energy-saving in large industries (Emission trading, 2021). Similar changes of emission prices are also observed in other countries. Comparison of the sectors within and outside the EU-ETS indicates that substantial emission reduction is achieved in the EU despite low prices of the allowances, but the impacts of ETS on renewable energy are not shown (Bayer & Aklin, 2020). Other assessments of the impact of allowance prices on particular renewable energy technologies are not found.

Presently, policies usually support fossil fuels while CO_2 pricing through taxes and emission trading is pursued to be developed. For CO_2 emission reduction, mainly the elimination of the subsidies for fossil fuels, combined with compensations for low incomes is the most effective and cheapest instrument in energy policies. The CO_2 cap with emission trading is proved to be reasonably effective but not as efficient as expected, but the grandfathering of emission is contentious because unfair to the firms that reduced CO_2 emissions ahead of the regulations and it impedes emission reduction.

5.4 Support for Renewable Energy

Policies can also encourage renewable energy through regulations public investments. Policies on renewable energy emerged in the 1970s in response to the oil embargoes imposed by oil producers on several countries, in particular the Netherlands, Portugal, South Africa, and the USA. These embargoes were imposed by the Arab countries in the Organisation of Petroleum Exporting Countries (OPEC) as sanctions for their support to Israel during the war of 1973, against Egypt and Syria, usually labelled as the 'Yom Kippur War'. As these sanctions addressed countries with various political settings at that time – the social-democratic government in the Netherlands, military in Portugal, apartheid in South Africa, Republican government in the USA – this action in 1974 shook the confidence in the global supplies of fossil fuels. Given that OPEC became a considerable global power, many countries pursued less dependence on the international trade of fossil fuels through a larger domestic production and storage, higher energy efficiency, and larger renewable energy. In the early 1970s, renewable energy referred mainly to the processing of domestic biomass for energy, hydropower, and geothermal energy in a few countries, whilst wind and solar energy were in infancy. The success of this boycott encouraged a coordinated limitation of the oil production by OPEC, which invoked high prices of oil and gas from 1979 to 1985, followed by inflation and economic stagnation in high-income countries. High oil and gas prices invoked policies on renewable energy in several countries. These policies were instrumental to the growth of modern renewable energy based on geothermal energy, wind energy, solar energy, and biofuels.

Policy support for renewable energy emerged in several countries largely shortly one after another. Brazil launched the PROALCOOL program, aiming at the substitution of oil imports for bio-alcohol based on its sugarcane processing. The USA launched the Public Utility Regulatory Policies Act (PURPA) in 1978 that promoted energy conservation and domestic energy resources including renewable energy, in particular blending gasoline with biofuels (Duffield & Collins, 2006). Denmark, Portugal, Germany, and a few other European countries also produced renewable energy from biomass by the end of the 1970s with supportive policies. That OPEC action dissolved in the mid-1980s when Soviet Union, Canada, Venezuela and other countries increased production of oil and gas followed by a sharp decline of the international prices. The formation of the European Union in 1993 by the Treaty of Maastricht consolidated the co-ordination of R&D on renewable energy into the hands of the European Commission (EC) whilst countries' authorities kept their own energy policies until the mid-1990s, when the EC launched its policies on renewable energy and energy efficiency including ETS mentioned above (Solorio & Bocquillon, 2017). Many countries introduced policies on renewable energy after signing of the Kyoto Protocol in December 1997, which included multilateral funding for renewable energy aiming to reduce CO_2 emissions in mid- and low income countries. As many countries GDP grew fast and many pursued had energy policies by the 2000s, the increasing fuel prices from 2005 onwards did not cause major economic shocks; instead, renewable energy is enhanced. During the 2000s, and 2010s, the number of countries with policies for renewable energy and mitigation of climate change tripled to nearly 150 countries of 196 countries that signed the Paris Agreement on mitigation of climate change. These are nearly all countries in the world (IRENA, 2018).

By the 2000s, various financial instruments for renewable energy were introduced. They are often called 'feed-in tariffs' but they are usually subsidies delivered to energy producers before supplies rather than price guarantees for the supplied services. Mixing subsidies and guarantees persists in the policies and in the OECD statistics on the 'feed-in tariffs' in the countries' policies (OECD, 2021). Despite the confusing definitions, it is assessed whether these 'feed-in tariffs' enhanced renewable energy. This OECD database covers feed-in tariffs for 7 energy resources in 36 OECD countries and 33 non-OECD countries during the period 2000–2019. While the feed-in tariffs during low fuel prices were rarely above a few dollars-cent per kWh, typical ranges during high fuels prices from 2005 to 2015 were for solar power 0.20–0.40 dollars-cent per kWh, wind power 0.05–0.11 dollars-cent, small-scale hydropower 0.02–0.08 dollars-cent, bioenergy 0.05–0.10 dollars-cent, energy from waste 0.02–0.08 dollars-cent, geothermal energy 0.05–0.19 dollars-cent and marine energy 0.04–0.10 dollars-cent per kWh. Thus, these feed-in tariffs varied per country and resource. The high ranges of feed-in tariffs were well above the high prices of oil and gas. As the policy support for renewable energy is rarely considered instrumental during low fuel prices, only 14 OECD countries and 4 non-OECD countries had feed-in tariffs before 2005, compared to 30 OECD and 23 non-OECD countries during 2005–2015. By then, the cheapest renewable energy based on hydropower and bioenergy was competitive with fossil fuels. The highest feed-in tariffs were for modern renewable energy because they were not competitive without such policy support even at the peak prices of oil and gas because, among others, the coal prices for electricity generation remained low. When the fuel prices decreased after 2015, not many countries increased the feed-in tariffs as a compensation of the decreasing fuel prices. Those that did increase their feed-in tariffs were a few EU countries, Japan, Switzerland and Turkey (among the OECD countries), as well as China, India and Indonesia (among the non-OECD countries). This illustrates that the countries' energy policies consider various issues aside from the prices of fossil fuels.

Regarding the policy support for modern renewable energy, a larger production can be expected. This expectation is assessed through correlations between the feed-in tariffs in the OECD database and the annual production of solar, wind, and geothermal energy combined with bioenergy in the BP database; the latter is used as a proxy for the feed-in tariffs on bioenergy. Other resources are ignored as

Table 5.1 Regressions of feed-in tariffs and renewable energy production in 57 countries based on the OECD patent data

Number of cases	Solar	Wind	Biomass	All
$R^2 > 0.6$	4	6	11	21
R^2 0 to 0.6	11	21	14	46
R^2 0 to – 0.6	24	14	11	49
$R^2 < -0.6$	3	3	1	7
All	42	44	37	123

definitions of small-scale hydropower and waste vary, and wave energy is small scale. This production data matched the feed-in tariffs in 57 countries, of which 36 were OECD countries. Combined with the resources, they represent 207 cases. Throughout the whole period of 2000–2019, the production of renewable energy increased in all those countries. While 127 cases of feed-in tariffs increased (39 biomass, 43 solar, 45 wind), 10 cases decreased (3 biomass, 3 solar, 4 wind) and in 70 cases the feed-in tariffs with nil (27 biomass, 23 solar, 20 wind). Table 5.1 summarises the results of correlations made on the assumption that the regressions (R^2) higher than 0.6 and lower than –0.6 indicate an important enhancement or impediment, whilst the regressions in-between indicate changes only in a few countries; a few feed-in tariffs with disputable data and nil tariffs are excluded.

Strikingly, 56 negative regressions of feed-in tariffs and renewable energy production were observed (all R^2 below 0), while only 10 of the negative signs can be explained by the decreasing feed-in tariffs. Possibly, those OECD data are deficient or subsidies for fossil fuels had a larger impact than the policy support for renewable energy. Most feed-in tariffs had moderate positive or negative effects. Only 4 cases in solar energy, 6 in wind energy, and 11 in bioenergy with geothermal energy had strong positive effects on their production; which is only 17% of the total number of cases. It is an unimpressive result of the policy support for renewable energy across countries throughout low and high prices of fossil fuels which indicates inconsistent policies. For example, substantial policy support is delivered by a few countries in the EU during high fuel prices, but this support is reduced during low fuel prices in the late 2010s, which is inconsistent with the idea of feed-in tariffs as an instrument that creates competitive edge for renewable energy.

The feed-in tariffs aim to invoke private investments in renewable energy while public investments also generate renewable energy. Based on the evolutionary string of thinking large public investments are needed because they generate renewable energy production and can invoke larger private investments through infrastructure as pipelines for residual heat and electricity networks. However, large public investments can also cause obstructions of the private investment when they are used for supplies of energy services delivered at prices below the market prices. The scale of all public investments is unclear. An authoritative BNEF database shows that the annual average of global investments in renewable energy was USD 210 billion from the year 2004 to 2017. This is about 15% of all global investments in energy based on the IEA data (IEA, 2021a, b). Those investments in renewable energy grew when the prices of fossil fuels increased from 2004 to 2011, which is followed by a saturation of the public investments at about USD 290 billion a year during the high fuel prices, and decline after 2015 when the fuel prices declined. The IRENA data show similar annual average investments of USD 301 from 2013 to 2018 (IRENA, 2021a). There exists a broad agreement that higher investments in renewable energy are needed for the energy transition, albeit the scale of investment is debated. A possibility to enhance those investments is through larger public investment.

With respect to the public investments, the BNEF refers to the public markets and governmental investments which show 8% share in all investments in renewable energy during 2004–2011. However, BNEF data does not specify whether the investors in assets are private or public organisation, though they cover 71% of annual average investments during that period. The IRENA shows that the public investments in renewable energy were 17% annual average during 2013–2018. An analysis of the BNEF data suggests that the share of public investments is 27% annual average investments in renewable energy from 2004 to 2014 (Mazzucato & Semeniuk, 2017); however, this cannot be verified because the website referred to is not reachable. Moreover, the BNEF data shows a decline of the public investments in total, the IRENA shows fluctuations (IRENA, 2021b), whereas that analysis shows a fast increase toward one-third in 2014. Given the variation in the estimates of annual average public investments from USD 20 billion to nearly USD 140 billion, and differences in the assessed trends a meaningful conclusion is impossible. While many countries' policies can agree on the importance of larger public investments for renewable energy production and infrastructure the crowding-out effects should not be expected regarding low share of the public investments in all investments in renewable energy. A contrary effect can be expected. Larger public investments are also important signs of confidence in modern renewable energy for the private investors, which is particularly important for, yet, costly distributed energy systems.

Many countries supported several resources in renewable energy with feed-in tariffs per energy unit which exceeded the market prices of fossil fuels. However, their impacts measured by larger production energy were modest. Larger public investments in renewable energy are needed and they are possible regarding its modest share in all investments.

5.5 Stakeholders in Renewable Energy

The idea that policies invoke innovations is disputable. For example, the Californian zero-emission vehicle regulation of the late 1990s did not invoke electric cars (Dixon et al., 2003), neither the Brazilian biofuels regulations in the 1990s replaced fossil fuels (Gasparatos et al., 2012). Whether the *Energiewende* in Germany is going to create a fossil-free economy, or generate more gas from abroad is to be seen during the 2020s. If far-reaching policy demands are posed, they based on novel technologies that are supplied by innovators after tests in practices but usually, the policy demands are tuned to the available technologies from the past. Fortunately for the energy access and mitigation of climate change, various renewable energy technologies are available from the late 1960s onward.

A behavioural idea about innovating is that innovators emerge spontaneously, driven by the interests to resolve problems, accrue incomes, 'do good', and other personal objectives. An innovator usually involves several partners and customers within a local community, which generates finances, know-how, and other resources. Such innovators' networks enable startups. In modern renewable energy, the motivations can be the creation of local businesses; for example, the bio-ethanol business from a surplus of sugarcane and energy production from wood residues. They also pursue local energy on isolated locations; for example, in isolated villages, and in space. Others aim at accessible energy for low-income, as solar cookers and lamps for households. Such innovator networks emerged in the late 1960s and early 1970s but many failed in business. A few successful ones in the niche markets during the 1980s generated large-scale activities and income; a large-scale refers to regular production. During subsequent decades, large-scale activities expanded. Herewith, policies supported such innovator networks with subsidies and regulations or obstructed them when gave support to the interest vested in the past. For example, the Danish innovators in the wind turbines in the early 1970s obtained the policy support after nearly ten years of deliberations, whilst the Dutch innovators were dismissed by the Dutch policy based on opinions of the large, incumbent corporation about the small market for the wind turbines in the future (Kamp, 2006); the Danish innovators became the global leader in wind turbines. Market prices were not decisive for the innovators as the early initiatives in modern renewable energy emerged during low prices of fossil fuels. High prices of fossil fuels during the 1980s enabled large-scale production, and triggered the interest of customers and policymakers, entailing growth of modern renewable energy during the low fuel prices of the 1990s.

Large-scale bioenergy and geothermal energy emerged in the late 1960s during low fuel prices and non-existing policies on renewable energy. Large-scale wind and solar projects above 1 MWh production started in the early 1980s during the high prices of fossil fuels while their costs per energy unit were several times higher than the peak prices of fossil fuels but subsidies were scarce. Table 5.2 shows the years and scale of early projects in modern renewable energy from 1965 to 1989 based on the BP and SHIFT databases.

Bioenergy and geothermal energy emerged in France, Italy, New Zealand, and the USA in the 1960s, followed by Brazil, Germany, and Japan in the early-1970s, as well as Mexico, Canada, and the Philippines from the mid-1970 on; by far the largest production was in the USA. Wind power emerged in Sweden during the early 1980s, followed by several countries during the late-1980s when the fuel prices were low. Solar energy production started in the USA during high fuel prices in the early-1980s, followed by the European countries in the late-1980s when the fuel prices were low. Large scale bio-alcohols and biooils started during low prices in the early-1990s in Brazil and the USA, followed by several European countries.

Table 5.2 Introduction of modern renewable energy; volume and years of initial production, based on BP and Shift data

	Geothermal with bioenergy (>1 TWh)		Wind energy		Solar energy		Biofuels	
All in GWh, PJ/GWh = 278	Started in	1965–89	Start in	1965–89	Start in	1965–89	Start in	1990–94
North and Latin America								
Brazil	1970	1505					1990	75,044
Canada	1976	1033	1985	0.2				
Mexico	1973	1146						
USA	1965	25,451	1983	86	1983	12	1990	23,852
Europe								
Austria							1990	82
Belgium			1987	1				
Denmark			1986	41				
France	1966	1351					1992	254
Germany	1970	1132	1986	1			1992	79
Greece			1987	0.1				
Italy	1965	3665	1989	0.1	1989	0.1		
Netherlands			1986	2				
Portugal			1989	0.04				
Spain			1989	1	1989	0.2		
Sweden			1983	1				
United Kingdom			1989	0.4	1988	0.2		
Asia and Oceania								
Japan	1970	4563						
New Zealand	1965	1573						
Philippines	1977	1925						
Average during 1965–1989								
BP data		135,396		390		36		0
SHIFT data		131,243		985		1698		64,554

By early 1990s, wind energy expanded. Three entrepreneurial initiatives in wind energy in Europe are illustrated in Table 5.3 based on interviews with key persons in these leading initiatives at that time in Europe.

These initiatives in the early 1990s were not taken by experts or businesses with a long track record in renewable energy, because the initiators were usually newcomers who learned in the practice. The backgrounds of these entrepreneurs varied. The Ecopower co-operation in Belgium was an initiative of activists against nuclear power; the founders were linguist specialists that started a firm with 30 participants which grew to be the largest energy co-operative in Europe with more than 60,000 members in 2020. On Samsø, a small island in Denmark with 4000 inhabitants, local farmers looking for business opportunities took a course on energy and started with a wind turbine. What followed was the growth in renewable energy for complete

Table 5.3 Drivers of early initiatives in renewable energy in Europe, summary of interviews

	Ecopower co-operative	Self-sufficient Samsø	Wind energy Navarra
Interview with initiators	Dirk Vansintjan and Relinde Baeten, co-founders, both Master Linguistics	Søren Hermansen co-founder farming course, course sustainable energy planning	Begoña Urien Angulo, capacity trainer, MBA and Ph.D. holder in psychology
Location and scale	Belgian cooperative, start with 30 cooperants in 1994, 60,000 in 2020	Danish island, start mid-1970s for 4500 inhabitants but 3700 in 2020	Navarra region in Spain start mid-1990s, 520,000 people, 640,000 in 2020
Main motive	Alternative for actions against nuclear power	Alternative for crisis on the island in 1990s	Use local resources by public utility, called EHN
How it started	Won tender for 3 mills due to social innovation	Awarded Masterplan for renewable energy islands	EHN-director invested in wind and small hydro
How income grew	Fair price for electricity service, sharing profits, good opinions of clients and experts; doubling membership after the Fukushima explosion	Policy support windmill; then, 11 wind mills, heat pumps, insultation, now self-sufficient, 5 offshore mills, biofuel, 60MW solar, Energy Academy	Power sales for region and export; joint EHN-Inberdola, expansion till early 2000s; then, taken-over by Gamesa-Vestas-Sodena and stagnation
What are main losses	Bankruptcy of the firm for balancing of power network	Technical university on the island unsuccessful	Stagnant network because no earning for community
What can be better	Wood pellet factory, finance & ecology	Local financial institution for small loans	Distribution and capability in the region
What is the result	2% of Belgium energy market	More than 100% energy self-sufficiency	2460 MW renewable 40% wind, 23% self-sufficient
Future challenges	Energy poverty, state guarantees to citizens and firms, centre for new co-operative ideas	Electric mobility, smart grid, storage, hydrogen, a local 'greenfund' and know-how	Electric mobility, smart grids, link businesses to communities' benefits

energy self-sufficiency on the island in the early-2010s. The initiative in the Navarra region of Spain was more professional as it was triggered by a local state energy firm in the early-1990s but not per se expert-based. On behalf of the regional government, it was coordinated from the late 1990s to the late 2000s by a social psychologist. By 2020, renewable energy covers about 20% of the regional energy consumption, the equivalent of 50,000 people. Motivations differed, whilst all initiatives used wind energy. While the Ecopower aimed at alternatives to nuclear power; the Samsø initiative was triggered by an economic crisis on the island, and Navarra pursued the development of local natural resources.

All these initiatives aimed at income generation rather than maximum profitability. The Ecopower grew due to good service and price which created positive opinions among customers and experts, the Samsø income was from sales of energy to its community with financial support from the Danish government, and income in

Navarra was based on energy sales by the regional public energy producer with support of national feed-in tariffs. In Navarra, the public producer was acquired by large foreign firms when wind energy was successful, the privatisation is enforced in Spain, and the wind energy production was enhanced by regulations. By 2020, Ecopower served 2% of the Belgian energy market, the Samsø initiative created a self-sufficient island. but the Navarra wind power stagnated after that acquisition despite high fuel prices in the late-2000s. Difficulties with the power networks were experienced because the municipalities in the region had no interest in constructing of infrastructure for the energy networks. Ecopower suffered when the firm that balances electricity on grid went bankrupt. Samsø had also difficulties in the upgrading of the network. All those entrepreneurs experienced impediments in creating of financial and human capabilities and expected to improve performance due to financial guarantees, local greenfunds and profit-sharing between business and community. Local energy networks with storage and electric mobility are perceived as key technologies of the future, and know-how about social participation, funding, and benefits is considered important for the development of these networks.

This summary of three early initiatives sheds light on the similarity between many stakeholders initiatives in the EU. Many initiatives emerged when several entrepreneurial persons pursued modern renewable energy in co-operation with communities while aiming to combine income with values as cleaner production, human scale, and other ethical qualities compounded in their energy services. This process of decision-making with the stakeholders' involvement, often branded as 'bottom-up', is called the stakeholder-based. Its rivals are the decision-making of large-scale corporations focused on profits, called market-based, and one of the governmental policies preoccupied with the political power, called policy-based (Walker & Devine-Wright, 2008). Furthermore, many initiatives deliberate between higher efficiency due to large-scale energy production, thereby larger subsidies, versus deliveries of energy services tuned to local demands of social and business networks (Krozer, 2012).

Cases in the EU indicate that stakeholders' initiatives often generate enterprises tuned to the local demands and interests (Seyfang & Haxeltine, 2012). Inquiries into the stakeholders' initiatives in Italy, Romania and the Netherlands show a mixture of social organisations, institutions and firms that usually pursue sufficient income for economic continuity, not maximising profits. The main success factors, as mentioned by the initiatives, are the motivation of participants and rooting in local communities; whereas poor economic performance is considered the main risk of the initiatives (Boon, 2012; Dragoman, 2014). Based on the cases in the Netherlands, it is argued that many stakeholders' initiatives are small-scale and keep operating in the market niches, thereby they do change dynamics in the energy markets (Arentsen & Bellekom, 2014). This assessment is valid for the situation in the Netherlands with supportive policies for fossil fuels. In Austria, Belgium, Denmark, Germany, and a few other European countries, the stakeholders initiatives turned into large-scale activities. However, information about continuity assessments of the stakeholders' initiatives in the competition is not found.

Early initiatives in renewable energy were pursued by innovators networks with various stakeholders during low and high fuel prices, followed by policies. These initiatives were driven by personal ambitions and ethical considerations rather than profit maximisation.

5.6 Start-Ups and Employment

A conventional argument is that the large scale of operations in modern renewable energy enabled lower costs, thereby higher performance of services in markets. This argument is used for justification of mergers between firms into the private conglomerates in various sectors of energy production. This is fostered by states on the arguments that the private suppliers perform better than the state firms, whereas the risks of monopoly prices would be countered by the regulations. A different argumentation is that high accessibility of renewable energy enables tuning to the local demands with the specialised services. For convenience, this ongoing debate about economic policies can be labelled as the economy of scale versus economy of scope. Herewith, it is not claimed to present a verdict, but underpin that the economy of scope in the EU has generated more startups and larger employment compared to the economy of scale in the USA. Based on these two indicators, larger social benefits are generated through the economy of scope.

A widespread opinion in the late 1900s was that public energy companies were poor performers. Therefore, they were privatised in many countries. A consequence was a large loss of employment from 1997 to 2017. Estimates are made based on the USA Census of establishments and EU statistics of firms (Eurostat) in the electricity and gas sectors. Employment in the USA energy sector declined by about −1% annual average and at a similar rate during most years throughout that period. In the EU, where three times more people worked, and a few times more people per establishment, employment in the energy sector declined by about −3% annual average, and that decline fluctuated strongly per year, which was problematic for the social arrangements aiming to find other jobs. Meanwhile, modern renewable energy was rarely embraced by the incumbent energy firms which operated based on fossil fuels. Besides these incumbents, value-driven innovators that involved stakeholders started production and service companies based on modern renewable energy. Most innovators were private firms. Widespread citizens' participation in renewable energy and energy efficiency is achieved in some countries. For instance, citizens in Germany possessed 47% of all renewable energy assets in 2015. That large share of all assets in renewable energy is based on numerous, small investments of many citizens (Yildiz, 2014). A number of regions in the EU also pursue suchlike citizens' participations with policy support (Interreg, 2021).

The impacts of those innovators on the birth of firms and additional employment in the USA and the EU during 2008–2016 are assessed based on those USA and EU statistical data. Figure 5.3 shows indexes of the firms' birth in all sectors – abbreviated as 'all firms' in the Figure – and in energy sectors in the USA and EU. The additional jobs are discussed below. The period from 2008 to 2016 covers high fuel

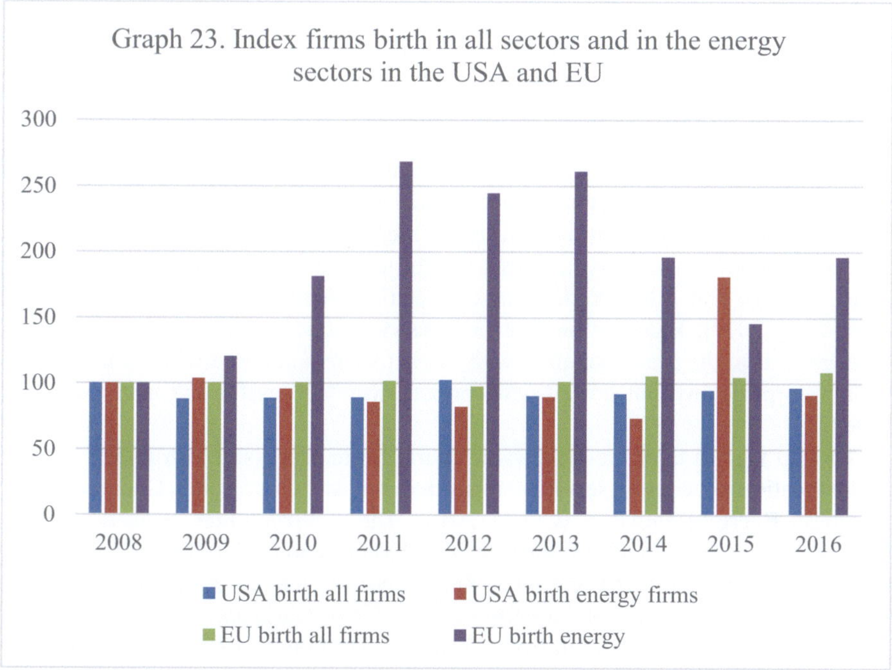

Fig. 5.3 Index firms birth in all sectors and in the energy sectors in the USA and EU

prices and large policy support for renewable energy driven by the political will to compensate incomes for losses after the financial collapse in 2008; the USA policy support was somewhat larger than the EU support. In this period, wind and solar energy grew particularly fast.

During the fast growth of modern renewable energy, the number of newly born firms in the USA energy sector grew similarly to all sectors by an annual average of 10,000 and 700,000, respectively. The annual average of additional jobs was nil in the energy sector, while 890 000 in all sectors, which indicates a jobless growth in the USA energy business. In the EU, the number of firm's birth in the energy sectors grew much higher than in all sectors. While the firms' birth in all sectors was continuously about 2.5 million a year during that period, the firms' births in the energy sector grew from 10,000 in 2008 to 23,000 in 2011, but gradually declined to 16,000 in 2016; the annual average was 16,000 new firms. The annual average additional jobs in the EU were 22,000 in the energy sector and 630,000 in all sectors. The energy innovators in the EU generated more firms and a larger number of jobs than in the USA, measured by the total number and compared to the growth of all sectors. For illustration, this growth in the number of energy firms in the EU was 26 times faster than in the ICT sector. The EU gained firms and employment from those values-driven innovators that were focused on the specific services for local demands rather than bulky deliveries, but were disapproved by the incumbents. They labelled as "walkers with broken laces" by the former Director Environment of Shell International.

New firms in the EU generated novel, value-adding energy services called 'Energy Service Companies', or ESCO (Bertoldi et al., 2014). Besides the energy supplies, new firms have added value when addressed specific demands (e.g., micro-cogenerators), technology developments (e.g., micro-grids), energy storage (e.g., heat and power reuse), energy systems (e.g., heat-balance systems), and other functional and ethical qualities perceived valuable in energy consumption. Business models for demand management are pursued as energy performance agreements (e.g., air conditioning), continuous improvements, (e.g., annual energy reduction), co-operative performance (e.g., district heating), flexible pricing (e.g., automation of consumption), and other models for cooperation in the value chain (Bertoldi & Boza-Kiss, 2017). Valorisation of energy services emerged as a relevant business strategy next to the pre-occupation with market share and cost-saving in large-scale energy corporations, but not all new services were successful and many start-ups failed.

The EU feed-in tariffs for renewable energy were successful instruments for this diversification of energy services, measured by accounts of the Council of the European Energy Regulators (CEER, 2018). These expenditures on feed-in tariffs grew steadily from about 19 billion euro in 2009 to about 58 billion euro in 2017; whilst the number of the EU countries that introduced feed-in tariffs increased from 14 to 26; it means all member countries except for Bulgaria and Slovakia. The total expenditure of a country varied from 22 million in Malta to 17 billion in Germany; which means 0.1% and 0.4% of their GDP. Despite large differences, the correlations cross countries between feed-in tariffs and annual birth of firms the energy sector were high for years from 2009 to 2017 ($R^2 > 0.8$), and moderate in 2010 and 2011 ($R^2 > 0.6$). This indicate effective expenditures, measured by the entry of new firms. The annual average public expenditures per new job of those new firms were below USD 4000 in Lithuania and Romania, and above USD 100,000 in Italy and Spain. Even the highest public expenditures per new job in those novel energy businesses were below average expenditures per job in all sectors. In most countries, the expenditures were below USD 30,000 a year. This public expenditure was far lower compared to public annual payments per created job in most EU countries.

Socially motivated initiatives based on cooperation with stakeholders with policy support in the EU evolved within 30 years into numerous firms in production of modern renewable energy. The support with feed-in tariffs was effective, measured by the startup of energy firms and these firms generated many jobs at moderate expenses per job.

5.7 Types of Decision-Making

As the innovator networks in renewable energy evolved into global commerce, wind energy expanded about 230 times, solar energy 660 times, and biofuels nearly 3800 times between 1990 and 2015. Meanwhile, geothermal with bioenergy grew only 8 times while hydropower grew twice, similar to global energy production. Wind

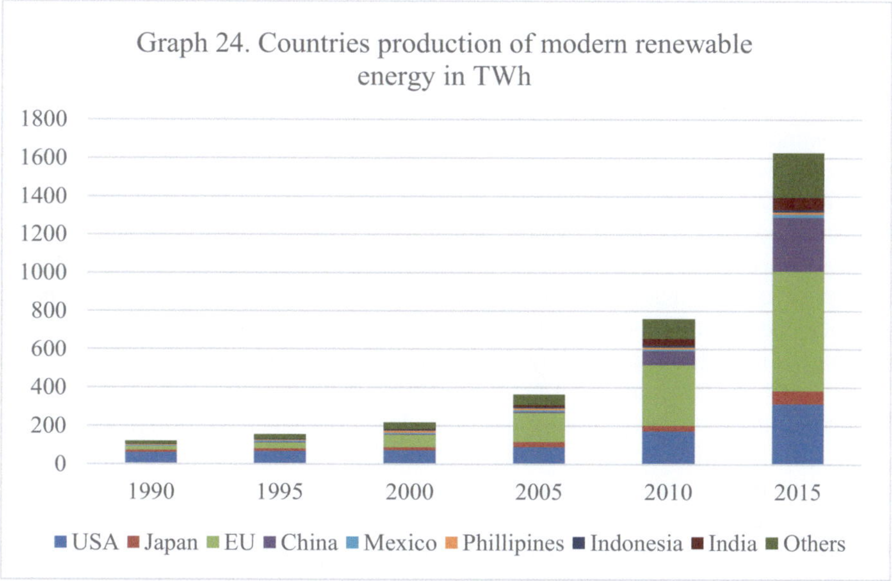

Fig. 5.4 Countries production of modern renewable energy in TWh

energy, solar energy, and biofuels grew fast during those 25 years into a key energy resource, despite their highest costs per energy unit within renewable energy whilst lower-cost hydropower, geothermal and bioenergy grew slower, but the OWD data indicates a decline of bioenergy.

That fast growth of modern renewable energy cannot be brushed aside as typical high growth rates of a novelty because fast growth rates continued for more than two decades, though the countries' growth rates varied. In effect, commercial activities in modern renewable energy turned into a global competition by the 2010s along with several leading countries. Figure 5.4 shows the production of modern renewable energy in eight countries during 1990–2015 in intervals of five years, based on the BP data.

By 2015, large producers in ascending order of their real income per capita were: USA, Japan, EU, Mexico, China, Indonesia, Philippines, and India. They covered about 85% of global production throughout that period, whilst 15% was covered by many countries labelled as 'Others'. Note that the countries' production was similar to consumption as trade with renewable energy was small scale. The shares of those countries changed as modern renewable energy grew globally by an 8% annual average during 1990–2015, compared to a 3% income growth and 2% energy consumption. In 1990, the USA lead with more than 50% of all modern renewable energy, but its share declined to 19% in 2015; meanwhile, the shares of China and India increased from nil to 17% and 4% while the EU took the global lead with 38% of global production in 2015. The remaining 22% of global production was divided between many countries with a particularly large share of geothermal energy in Indonesia and on the Philippines due to volcanic activities.

The reason for the shrinking share of the USA and Japan, which were the frontrunners in modern renewable energy during the 1990s, contrasts with forecasts of prominent energy consultants, during the early 2000s, who assessed stronger businesses in the USA and Japan than in the EU and China (Asplund, 2008; Pernick & Wilder, 2008; Siegel, 2008). For example, it is assessed that the USA's wind industries covered about 50% of the global wind power in the 1990s, and its firms were assumed to keep leadership, but higher share on the global market was captured by Danish and German firms a few decades later. Japan was leading in photovoltaics for solar energy in the 1990s, and the innovative superiority of Japanese electronical and mechanical firms was widely (Kodama, 1995). However, Japan lost a market share in PV to China and the EU. A plausible explanation for their decline is the decreasing interest of large companies in the USA and Japan when high aspirations of profitability were not met during the low fuel prices of the 1990s; the USA focused on oil and gas, Japan on nuclear power.

Those changes in the market shares are consequences of decision-making in various – novel and vested – interests in businesses, policies, and civil society. For convenience, the decision-making about the development of modern renewable energy across countries can be classified into three types. In market-based decision-making, large companies determine the course of action. Private companies decide about investments in innovations with the policy support for individual firms. These companies can be the newcomers that generate private funding and subsidies, as well as the incumbent firms that do not innovate themselves, but acquire innovators when they participate in new ventures. The economies of the USA, Japan, Mexico, and the Philippines are illustrative of this type of decision-making. In policy-based decision-making, the governmental authority is. The authorities decide about plans and investments in all kinds of energy mainly based on the state budgets, taking into consideration the rival interests of various businesses. The political decisions can support or impede modern renewable energy with respect to that consideration. This is typical for decision-making in China and Indonesia, where the state-owned energy corporations decide about investments within the framework of national plans. The stakeholders-based decision-making links private and public interests. Investments in modern renewable energy are made by private organisations with respect to the interests of communities, businesses, and civic organisations because the authorities and the interest groups have a say in the regulations and subsidisation of the economic activities. Among those countries, the EU and India are illustrative of this type of decision-making. The unclassified countries are compiled as 'Others'. This way, the types of the decision-making can be related to the global market shares in the production of modern renewable energy.

The shares of the market-based, policy-based and stakeholders-based decision-making in the global production of modern renewable energy from 1990 to 2015 are plotted in Fig. 5.5.

It is hypothesised that the countries which operated with market-based decision-making were fast in the initiating large-scale production of modern renewable

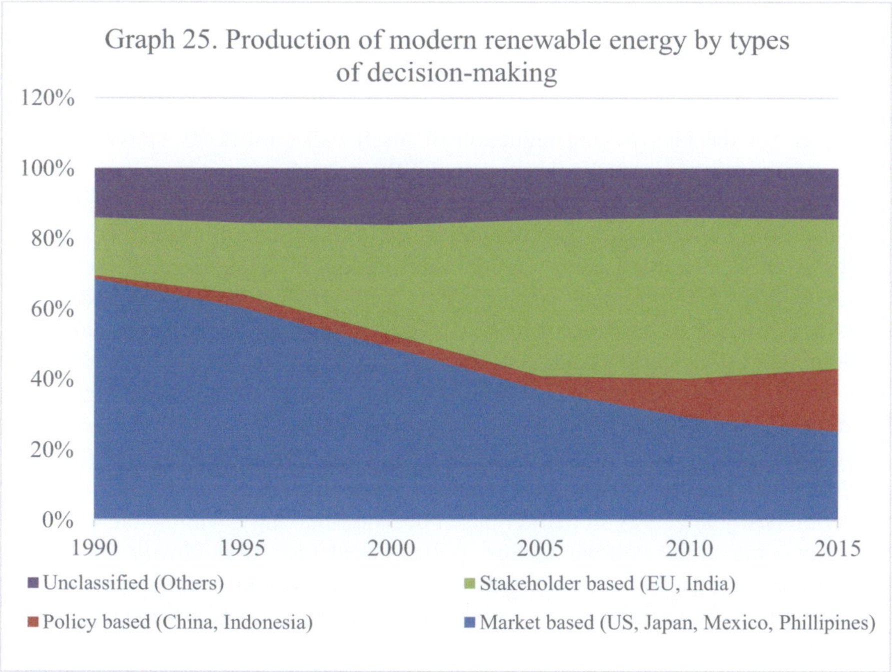

Fig. 5.5 Production of modern renewable energy by types of decision-making

energy. As the interests of large-scale firms declined because faced decreasing profitability during low fuel prices, the investor risks increased and investments in production declined, entailing a declining global market share of these countries. Meanwhile, countries with policy-based decision-making were slow in the introduction of investments in modern renewable energy because this involved laborious administrative process. Once the policy was approved, the investments were generated largely irrespective of the prices and particular private interests. Meanwhile, the modern renewable energy grew steadily in the countries with stakeholders-based decision-making because stakeholders combined private and public capital for investments in conformity with governmental plans. Such combinations could be arranged because the stakeholders' participation was instrumental in the policy support which reduced the risks of the private investments, thereby attracted investors.

Such types of decision-making about modern renewable energy are found not only in the front-running countries that are mentioned above but also in the lagging countries. For example, countries in the Middle East and North Africa (MENA) are endowed with excellent natural resources for solar power, as well as sufficient financial and human capabilities for the production of modern renewable energy. However, their production started only in the 2010s, as the BP and IEA data show. In Morocco, market-based decision making was led by a consortium of corporations in a large-scale project on solar power called 'Desertec', but the project failed to produce solar energy until the late 2010s on account of the failure to finance infrastructure and avoid costly corruption. Policy-based decision-making in the United Arab Emirates (UAE) pursued solar power, which was framed by the long-term plan for the far-reaching reduction of fossil fuel until 2050. However, large investments in concentrated solar power (CSP) in 2016 generated only 1 TWh of solar power in 2018; inter alia, its ambitious all renewable energy Masdar City failed (Masdar, 2021). Meanwhile, stakeholders-based decision-making, which is arranged by the Arab Renewable Energy Commission in Jordan, has generated more solar power in 2018 through small-scale projects with small firms and limited policy support compared to Morocco and UAE. Stakeholders-based decision-making in modern renewable energy is sluggish because various interests must be aligned, but it is robust when stakeholders are involved. This type of decision-making can be instrumental for countries endowed with natural resources for modern renewable energy, yet missing large human capabilities and financial means for that growth.

Countries' decision-making about modern renewable energy vary. The growth of modern renewable energy in market-based decision-making is vulnerable to changing prices and other external factors, policy-based one starts slowly because of administrative processes and can generate large investments, whilst stakeholder-based one is sluggish and grows steadily due to compounding of the stakeholders' interests.

5.8 Social Acceptance and Benefits

An obvious question is why people purchase modern renewable energy if costly and intermittent. A plausible answer is that it matches the demands of some consumers because delivers valuable qualities for energy consumption. It is a challenging issue to assess the consumer acceptance of particular qualities with regard to higher prices of renewable energy because the stated preferences usually differ from the preferences revealed in purchases. Despite observed differences, the conventional method is the estimation of the stated preferences about willingness to pay with the use of inquiries. As the statements do not deliver robust findings, but mixed expressions of consumers which depend on the socio-economic condition and the assessment method, a brief review of the willingness to pay is followed by indications of the revealed preference in purchases. More systematic assessments of the revealed acceptance of modern renewable energy would benefit decision-making about investments and policies but, unfortunately, not many estimates are found, let alone reliable statistical data.

The studies on the willingness to pay for renewable energy show ambiguous results. High willingness to pay is observed for electricity in rural areas in low-income countries with limited electricity infrastructure (IEG, 2008), as well as for lighting, education, and entertainment based on a grid (Wilson et al., 2010), whilst off-grid solutions are costlier than on-grid. Whether costlier off-grid solutions are acceptable is uncertain because the inquiries in high-income countries indicate a higher willingness to pay for the electricity on-grid than off-grid (Scarpa & Willis, 2010). Studies on renewable energy also show mixed results. For instance, a study in the USA revealed higher household payments for renewable energy than the willingness to pay stated in inquiries (Farhar, 1999). Opposite results are obtained in Korea, where the cheapest alternative is preferred but the statements are not verified with purchases (Kim, et al., 2013). Literature reviews show ranges of acceptable payments from a few dollars-cent to a few dozen dollars-cent per kWh (Streimikiene et al., 2019). One review shows the range of 100 times (Zorić and Hrovatin, 2012), but another one the range of USD 10–37 per month (Grilli, 2018), which is equivalent of 3–11 dollars-cent per kWh assuming 300 kWh a month. There is no agreement about the causes of those differences as some scholars have pinpointed at different inquiry methods (Ma et al., 2015), but others at varying income, education, age and other social factors in the assessments in high-income countries (e.g., Ntanos et al., 2018), and in low-income ones (e.g., Entele, 2020).

A cautious estimate of the willingness to pay of many households could be 5–10 dollars-cent additional payment for the renewable energy power if it is a reliable supplier, and if the taxes on electricity are not high. However, not much is found about the willingness to pay for renewable energy in heating, neither in small and large businesses nor for various modes of transport, which are major omissions as power at households covers only a part of all energy consumption. Below, the revealed preferences for modern renewable energy are addressed.

While early innovations in wind and solar energy were often driven by the innovators' interests in pursuing technologies with high ethical qualities but limited consumer application as wind turbines in remote communities, main consumers used functional qualities in the exploration of space and telecommunication. As those applications are scale up by early 1980s, modern renewable energy was valuable in farming, housing, transportation, telecommunication, aerospace and other services without direct connections to the networks of power and fuels. By 1980s, potential demand for wind and solar energy on those market niches was already substantially larger than the production of than time. For example, assuming that 5% of farmers in the USA had to use local energy resources because they were poorly connected to grid, their demand for the distributed energy systems approximated 60 TWh which was twice higher than all 28 TWh production of modern renewable energy that time in the USA; the USA farmers consumed 2.4 Quad BTU, i.e., 703 TWh in 1980 but data on renewable energy are not shown (Miranowski, 2005).

Less dependence from energy imports was a driver for the development of modern renewable energy after the oil embargo in the 1970s, as policies in the USA, Europe, and other countries pursued lower import-dependence. This justified public investments and support of private investments in nuclear power, tar oil, shale gas, as well as modern renewable energy. However, the impact of policies was modest. Despite the public appeal, these policies failed as the global trade grew continuously from 1980 to 2015 and the share of global energy imports in global energy production increased in all energy resources, particularly fast in biofuels and waste. High oil prices had more of an impact. While global oil production decreased by −1% annual average during the high oil and gas prices of the early 1980s, the imports decreased by −5% annual average but grew faster than production during the subsequent 20 years of low prices. In the USA and Europe, the imports declined even as much as −6% annual average during the period of high prices, but increased just as fast during low prices, particularly fast in the USA. The main impact of those policies was on the dissemination of modern renewable energy because domestic production could expand on a side of a lower consumption of fossil fuels during high prices.

During the late 1980s and the 1990s, larger investments in modern renewable energy generated cost-reducing technical change, while environmental issues gained attention. Mitigation of climate change was issued in the USA, followed by Europe (Dyke et al., 2021); whereas, accidents with nuclear power in Le Havre (France), Harrisburg (USA) and Chernobyl (Soviet Union) prompted the search for safer energy resources. Such concerns during the times of low fuel prices enhanced the policies for modernisation of energy industries in Europe wherein several countries pursued modern renewable energy for economic development. That choice was successful as novel industries with global technology exports are developed. By the early 2000s, the unit costs of wind power on land approached the price parity on grid because of costlier fossil fuels, which attracted large-scale investments. By then, the price benefits of modern renewable energy for individual producers and consumers were observed; for example, lower spot prices on electricity markets when trade with renewable energy was involved (Ciarreta et al., 2014; Gianfreda et al., 2016), and the income growth due to the production of renewable energy (Bhattacharya et al. 2016).

By the 2010s, benefits of renewable energy refer to the avoidance of costs due to better planning, deferred investments in the large-scale capacities, reduction of energy losses, as well as better management of ancillary costs, reliability, and fuels. Health and environmental benefits as a result of lower pollution, as well as economic benefits of less administration and transfers of energy along with larger employment, income, and capabilities are pinpointed. Appendix 4 shows the main economic benefits of renewable energy and energy storage based on literature. Though some benefits are anecdotal they can be used as a checklist for the integration of private and collective interests in investment decisions which enables to anticipate rewards and risks in projects, referred to as 'social return on investment' (Krlev et al., 2013).

Various private and societal benefits are generated due to growing investments in modern renewable energy. Integral decision-making with regard to these benefits enhances socially beneficial shifts in energy consumption.

5.9 Cost-Reducing Change

Despite the fast consumption growth of modern renewable energy, wind and solar energy, in particular, nearly all renewable energy consumption is based on biomass and hydropower. As shown in Table 1.2 based on the IEA data, global consumption of the geothermal, wind, and solar energy was about 2.3 PWh compared to 19.3 PWh bioenergy and hydropower in 2015. A question is why did wind energy and solar energy grow faster than other renewable energy resources despite higher costs and intermittent properties. An answer is relevant because could indicate possibilities for generating faster growth of all renewable energy, which would accelerate the transition from fossil fuels to renewable energy. Fast growth of a novelty could not be a sufficient answer as wind and solar energy grew faster during several years of high fuel prices in the 1980s, as well as subsequent 15 years of low prices until 2005 and low policy support, and 10 years of high prices and large policy support until 2015, and a few years of low prices, thereafter. The drivers could be exogenous to the energy market, which means not directly dependent on behaviour of the individual interests on this market – growth of population, prices, policies, and suchlike – and endogenous because driven by the behaviour of the market interests – technological novelties, mergers for the economy of scale, qualities for the economy of scope, and similar. Pinpointing at the main drivers enables focus in decision-making on the particular mechanism of change.

Given the population and income growth, the main external factors could be fuel prices and policies. High fuel prices would accelerate the growth of modern renewable energy, but this should have had a similar impact on all renewable energy resources. Given the global increase of fuel prices, while biomass was the most traded renewable energy resource, one could expect the highest acceleration in bioenergy. This is partially observed as high growth rates of biofuels but the volume of biofuel has been only 0.01 PWh in 2015 based on IEA data, though BP data shows manifold larger biofuel production. Larger policy support for wind and solar energy could be relevant. While the policy support in a few trailblazing countries was relevant during low fuel price during the 1990s, high feed-in tariffs were introduced during high fuel prices 2005–2015 when large-scale production in wind and solar energy was already realized. Furthermore, the OECD data shows that feed-tariffs for bioenergy, small scale hydropower, and geothermal energy, as well as for wind energy were all above USD 0.1 per kWh, and they were similar in many countries. Solar energy had a larger support per kWh and did not grow faster than wind energy based on the IEA data, but BP data show the opposite. Nevertheless, costly, intermittent solar and wind energy grew faster than comparatively cheaper, continuous energy resources. Therefore, these exogenous factors had a role but do not explain the faster growth of wind and solar energy compared to other renewable energy resources.

Various specific endogenous factors can be assessed which, in effect, have generated larger investments. Herewith, the growing investments are considered as indicators of the endogenous factors. In line with the evolutionary argumentation, investments in technologies could drive the growth of wind and solar energy. It is hypothesised that large investments invoke larger capacity because the investment cost decline as the volume grows (economy of scale), and even larger energy production because larger production enables higher capacity utilization (effect-increasing change) which generates the decreasing costs per capacity and per energy unit (cost-reducing change). In this reasoning, larger demands for wind and solar energy due to private and public interests and policies invoked larger investments which generated larger capacity and even larger scale of production, thereby fast effect-increasing technical change entails even faster cost-reducing technical changes. The increasing investments generated a faster decrease of the unit costs in wind and solar energy which fostered diffusion of this modern renewable energy. Global data are used to underpin this hypothesis. Firstly, the capacity and production growth associated with the investments are estimated; second, the effect-increasing and cost-reducing change associated with that production growth are assessed. It is called 'associated' in the sense of links between the scale and costs because reliable correlations cannot be assessed with this data.

Investments in the production of hydropower, bioenergy, geothermal energy, wind energy, solar energy, and marine energy during 2004–2017 are based on BNEF data. The installed capacity in GW and the production in TWh of those resources are based on the IRENA database. The IRENA databases also shown investment during 2013–2018, but the growth rates are higher compared to the BNEF database, except for bioenergy. More cautious estimates of BNEF are used in this assessment. If all investments, capacities and production put together are considered, larger investments are associated with higher capacities, and production. This expectation is plausible based on economy of scale. The investments based on BNEF data increased sevenfold from USD 41 billion in 2004 to USD 280 billion in 2017. Meanwhile, the installed capacities grew fourfold from 1518 GW to 5491 GW, and production doubled from 4962 TWh to 9007 TWh, based on IRENA data. Given the assumption about economies of scale in investments, this peculiar outcome needs explanation; assuming that those BNEF and IRENA data are correct, whereas using BNEF data for the investments does not change this result. Interpretation of these data could bring into conclusion that the investments in renewable energy are inefficiently allocated because the largest possible capacities with maximum utilisation of that capacity are not attained. This implies costly mitigating climate change.

An explanation based on the specification of technology data is that decision-makers invest in wind and solar energy rather than in larger-scale bioenergy and hydropower. Table 5.4 shows annual average investments in USD billions, capacities in GW, and production in TWh during 2004–2017, as well as the annual average growth during that period. It is done for hydropower, bioenergy, geothermal, wind, solar, and marine energy based on BNEF investment and IRENA energy data.

Global investments in hydropower and geothermal power were low and grew slowly, which is associated with slower growth of capacity, and production compared to investment growth. Investments in bioenergy were higher, but they hardly grew. Its capacity was larger, but this production grew slowly. Given that the capacities of hydropower and bioenergy covered nearly 89% of all capacities in renewable energy they determined the aggregated global outcome shown above. Meanwhile, larger and fast-growing investments in wind and solar energy generated faster growth of these capacities and even faster production growth as predicted

Table 5.4 Investments, capacities and production in renewable energy 2004–2017 based on BNEF and IRENA data

	Investment in USD billion		Capacity in GW		Production in TWh	
Average and growth per year	Average	Growth	Average	Growth	Average	Growth
Hydropower	6	10%	949	3%	3516	3%
Bioenergy	24	2%	1871	14%	622	8%
Geothermal	2	9%	10	3%	71	3%
Wind	80	16%	232	20%	521	22%
Solar	97	27%	108	45%	144	47%
Marine	0	52%	0	8%	1	7%
Total	210	17%	3170	11%	7975	4%

based on the evolutionary theory about economies of scale. Investments in marine technologies were low and grew fast, but the capacity and production grew slowly. Apparently, investments in hydropower, geothermal power and bioenergy were insufficiently beneficial. Low level of investments in those renewable resources impedes the accessibility of energy and the mitigation of climate change as their global volume is manifold larger than the volume of wind and solar energy.

Whether larger investments increase the capacity utilisation, thereby, decrease the production costs per capacity and per energy unit, is assessed with the IRENA data. The data cover the utilisation of installed capacity and unit costs per produced energy. These data refer to the IRENA estimates of the Levelized Costs of Energy (LCOE) of technologies. The IRENA data differ slightly from the LAZARD and NREL data, but the ranking and the proportion between the LCOE's of technologies are similar; recall that the LCOE is the unit cost estimated with standardised variables for investments and operations. Figure 5.6 shows annual average changes in annual production (in kWh), factor in the capacity utilisation (% of installed capacity), and cost reduction per capacity (in USD per kW) and per production unit (in USD per kWh) from 2010 to 2019 in percentages.

Low investments in hydropower, bioenergy, and geothermal energy were associated with low production growth, which in turn was associated with slowly growing capacity-utilisation, but the increasing costs per capacity and higher costs per production unit; the capacity utilisation even decreased in geothermal energy. It confirms the assumption about low-benefits of investment in bioenergy, hydropower, and geothermal energy. High investments in wind are associated with fast growth of the onshore wind and offshore wind, as well as improvement in utilisation of capacities and the reduction of unit costs. Faster investment growth of solar energy compared to wind is associated with production growth of PV and CSP. The utilisation of capacity also grows along with faster cost-reducing technical change. Thus, the hypothesis that larger investments in renewable energy were important drivers of the cost-reducing technical change in wind and solar energy is supported. Larger investments in wind and solar energy could overcome low fuel prices and decreasing policies of renewable energy after 2015 because generate fast

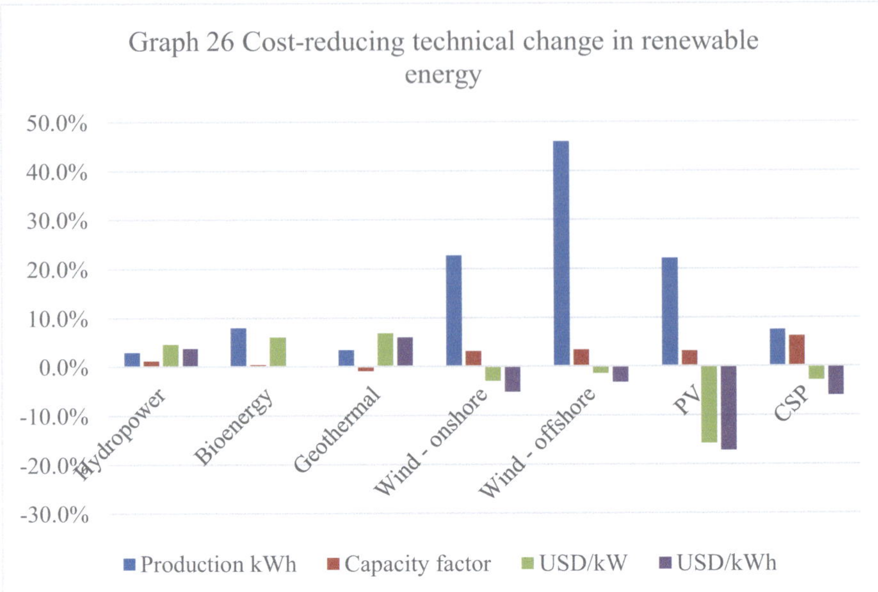

Fig. 5.6 Cost-reducing technical change in renewable energy

cost-reducing technical change. Higher investments can also explain the growth of wind and solar energy during the 1990s when fuel prices were low and policy support of renewable energy was limited. Hypothetically, higher investments in hydro, bioenergy and geothermal energy could also generate fast cost-reducing change but it is challenging to find beneficial investment opportunities.

A question is whether the cost-reducing technical change can go on as long as the investments grow. An answer is relevant for if the cost reduction of wind and solar energy can continue during several decades, these technologies enable growing access to energy and mitigation of climate change. Otherwise, investments in other renewables are necessary. For answering this question, it is estimated if the unit costs in USD_{2019} per kWh have declined over time. The unit costs of renewable energy cover the period 2010–2019 based on the IRENA data. Figure 5.7 shows the weighted average LCOE per kWh of energy resources estimated by IRENA; a few data on geothermal energy are interpolated.

The unit costs of all renewable energy resources can be compared to hydropower that is usually considered the cheapest renewable energy. Although the installation of hydropower is costly because USD 1000–7650 per kW of large-scale hydropower the operations cost only 2% – 2.5% of the installation which results in USD 0.02–0.19 per kWh; note that the installation costs of small scale hydropower are USD 1300–8000 per kW but the unit costs of USD 0.02–0.10 per kWh are not per se higher (IRENA secretariat, 2012). In these time series constructed with the IRENA data, the unit costs per kWh of hydropower were 4 dollars-cent in 2010 and increased during that very decade, whilst the unit costs of bioenergy and geothermal

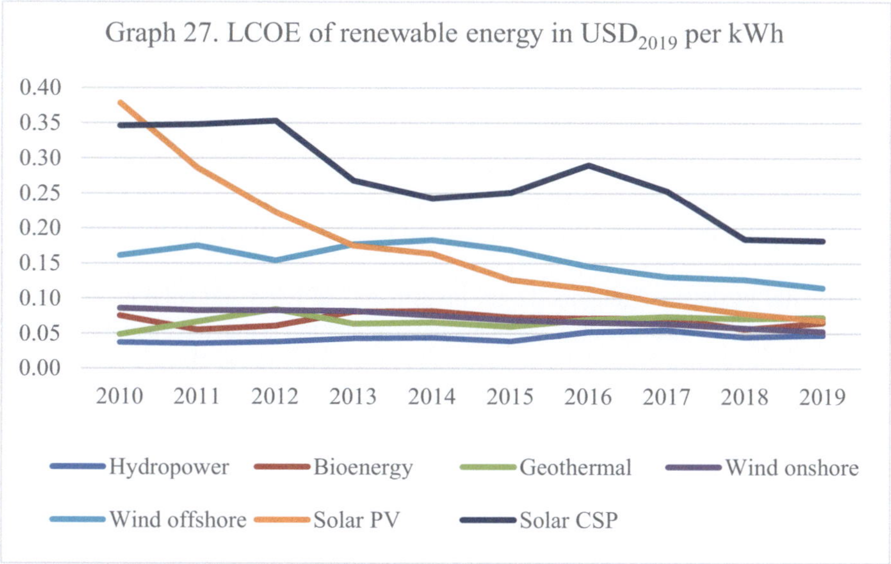

Fig. 5.7 LCOE of renewable energy in USD_{2019} per kWh

energy fluctuated between 5 and 8 dollars-cent per kWh. Meanwhile, the unit costs of onshore wind decreased from 9 to 5 dollars-cent per kWh, and offshore wind power from 16 to 12 dollars-cent per kWh. The solar Photovoltaics (PV) decreased faster as from 38 to 7 dollars-cent per kWh, and the Concentrated Solar Power (CSP) was slower as from 35 dollars-cent to 18 dollars-cent. All unit costs converged toward 5–7 dollars-cent per kWh. The lowest unit costs are within the range of unit costs of electricity production with coal power plants.

Those LCOE time series based on the IRENA data indicate that the unit costs converge across resources over time, which is shown in the Lazard and NREL databases. Nevertheless, a further decrease is possible. For example, the winning tender for CSP installation in the UAE promised a unit cost below 3 dollars-cent per kWh, and the tender for offshore wind in the Netherlands promised about 2.5 dollars-cent per kWh, though their performances after the realisation remain uncertain. Assessments showing saturation of the cost-reducing technical change were often mistaken. For example, an authoritative report on Danish wind energy that showed a −7% annual average cost-reducing technical change of onshore wind at suitable locations during the 1990s assumed a saturation by 1997 (Krohn & Morthorst, 2009), but the IRENA data shows a −5% change 20 years later. In the PV production in Japan, the cost-reducing technological change of −24% during 1974–1985 would slow down to −5% during subsequent 15 years (based on Watanabe et al., 2002), but the IRENA data shows a −15% of the PV applications 20 years later. Presumably, the cost-reducing technical change stagnates when R&D is small-scale or ineffective, and the applications are large-scale but poorly tuned to the local consumers' demands and environmental conditions. Otherwise, the cost reduction can continue for a long time. For example, when that PV production based on silica matured by the 1990s the cost-reducing technical change slowed down, but innovations in the organic, perovskite, and other semiconductors generated further cost-reduction. Therefore, it can be argued that more R&D and better application could generate beneficial investments in hydropower, geothermal energy, and bioenergy, which in turn, would generate faster cost-reducing technical changes, thereby expand the accessibility of energy and accelerate the mitigation of climate change.

Why did businesses generate larger investments in wind and solar energy compared to other renewable energy? A plausible explanation is that the business income grew from the growing sales due to fast cost-reducing technical change of wind and solar energy, though some supplying firms suffered losses when did not align to the pace of cost reduction. Comparison of solar energy, wind energy and hydropower in the USA illustrates these changes. Table 5.5 shows sales volumes of these energy resources and their unit costs derived from data on the USA business sales in USD_{2005} per kWh over time during 2005–2015 based on a private database (Statista Solar power, 2021).

The sales of wind and solar energy expanded from market niches attuned to the demands of specific customers towards mass markets, though the scale of traditional renewable energy indicated by hyrdropower was larger and hardly grew. In effect, the sales of solar and wind energy grew faster than hydropower despite initially higher unit costs; tapping into market niches that pay high price is essential in the early stages. This data source suggests that the unit costs of solar power were USD 2.67 per kWh in 2005. That time, it was more than 50 times costlier than wind power that was twice as expensive as hydropower, which was at parity with the lowest prices on the grid. During ten years from 2005 to 2015, the sales of solar power grew at an average of 22% a year whilst the unit costs declined by −48% ($R^2 = -0.6$), the sales of wind energy grew 17% and unit costs declined −6% ($R^2 = -0.9$), whilst the

Table 5.5 Sales, production and unit costs of the USA solar energy, wind energy and hydropower based on Statista Solar (2021)

	Solar energy supply			Wind energy supply			Hydropower supply		
	USD million	MWh produced	USD /kWh Cost	USD million	MWh produced	USD /kWh Cost	USD million	MWh produced	USD /kWh cost
2007	43	16	2.675	1783	34,450	0.052	6673	247,510	0.027
2008	47	76	0.614	2718	55,363	0.049	6542	254,831	0.026
2009	54	157	0.343	2965	73,886	0.040	6059	273,445	0.022
2010	69	423	0.163	3465	94,652	0.037	6037	260,203	0.023
2011	125	1012	0.124	3997	120,177	0.033	6065	319,355	0.019
2012	150	3451	0.043	4301	140,822	0.031	6119	276,240	0.022
2013	165	8121	0.020	4691	167,840	0.028	6258	268,565	0.023
2014	172	15,250	0.011	5213	181,655	0.029	6389	259,367	0.025
2015	189	21,666	0.009	5767	190,719	0.030	6540	249,080	0.026

sales of hydropower declined −0.2% and unit costs grew 0.3% ($R^2 = -0.9$). As high unit costs did not impede sales, as some customers obtained benefits, initially costly supplies out-competed the cheaper suppliers when could reduce their unit costs. Although some individual firms suffered losses because of fast cost reduction, the energy services generated larger incomes from the growing sales despite a fast decrease of the unit costs. This observation underpins the idea about innovations as processes of compounding qualities valued by the customers in niche markets at a high price, followed by the cost-reducing technical changes that enable the growth of sales as predicted in the behavioural theory.

It is underpinned that the faster growth of wind and solar energy compared to other renewable energy resources is driven by larger investments that trigger the effect-increasing and cost-reducing technical changes. As the cost reduction enables sales beyond market niches, energy services generate higher income than larger scale, but stagnant businesses which confirms the viewpoint that costly innovations expand when deliver qualities entailing cost-reducing adaptations.

5.10 Conclusions

A question is answered as to why renewable energy grew. Accordingly, the answer is focused on modern renewable energy based on biofuels, geothermal, wind and solar energy because they grew fast during the last 50 years despite high costs; whilst the traditional renewable energy based on biomass, waste and hydropower grew slowly. Answers are discussed from the mainstream perspectives focused on prices, institutional viewpoint on public policy, and behavioural approach that addresses values of energy services, followed by estimates of the benefits and cost reduction.

The mainstream explanation that innovations emerge as a result of price competition with the incumbent technologies is valid only for the early initiatives. Oil embargos followed by high oil and gas prices from the early-1970s to mid-1980s invoked the large-scale production of modern renewable energy. However, its growth continued throughout the 50 years of low and high prices. While the growth rates were higher during the high price years between 2005 and 2015 the changes of prices and volumes were unrelated, except for geothermal energy that grew slowly. Another popular explanation is that policy support enabled the growth of modern renewable energy. However, the support for fossil fuels was larger and more widespread across countries. Meanwhile, the support for modern renewable energy was incidental and limited to a few countries during the years of low fuel prices from 1985 to 2005. During the high fuel price years from 2005 to 2015, policy support increased and disseminated across countries; however, historical and cross-countries correlations of the production and support were rarely high. More effective for CO_2 reduction was cap with emission trading in the EU but it is unclear whether this policy contributed to the growth of renewable energy, or energy-efficiency, changes in composition of energy resources, shifts to unregulated areas called 'outshoring', and other factors in emission reduction.

Of greater importance for the growth during 1990s and 2000s was that entrepreneurial initiatives combined societal interests in renewable energy values in niche markets with income generation so as to generate business tuned to the communities' demands. They mobilised stakeholders for the purchase of their services and policy support and suitable policies for that business. In the EU, numerous new energy firms emerged that created jobs at low costs per job and valuable energy services entailing large scale activities in renewable energy. High prices of fossil fuels and policy support were instrumental for initial activities whilst the capabilities to generate customers, investors and financial support were decisive for the growth during 1990–2015. Joint capabilities of stakeholders explain faster growth of modern renewable energy in the EU compared to the market-based decision-making in Japan and the USA though the policy-based decision-making in China and Indonesia catch-up. The stakeholder-based decision-making is a useful model for countries that aim to generate the frontrunning business.

Larger investments in wind and solar energy were attractive because generated private and societal benefits due to the functional and ethical qualities as perceived by consumers. While in the early-1970s those benefits of wind and solar energy refer mainly to the early adopters in remote areas and military uses, from the early 1980s onwards, it is when large scale production started, more beneficial applications are found whilst the cost of deliveries declined. As the production scale increased, several countries that struggled with the economic crisis of the 1980s pursued policies on industrial modernisation through public and private investments in modern renewable energy. They generated the business start-ups, technology exports and creation of jobs in the 2000s when fuel prices increased, while the unit costs of renewable energy decreased. Meanwhile, several beneficial attributes for the individual and societal interests of producers and consumers are compiled into a checklist, potentially useful for decision-making in those countries and businesses that aim to stay abreast in this market.

Within modern renewable energy itself, wind and solar energy grew faster from the 1990s onward than bioenergy and hydropower that were manifold larger by scale and cheaper. This peculiarity from the mainstream viewpoint can be explained from the evolutionary perspective. The explanation being that larger investments enabled cost-reducing technological change because they generated economies of scale in the manufacturing of equipment, and better utilisation of applications. This explanation is underpinned with the global data on investment, capacities and production and from 2004 onwards and data on the decreasing unit costs of technologies during the 2010s. As a result of larger investments, the costs per unit of installed capacities of wind and solar energy decreased, whilst they increased in other renewable energy energy resources. Low level of investments impedes the accessibility of energy and the mitigation of climate change because bioenergy and hydropower still covers about 90% of all renewable energy.

Those estimates with global investment and production data are confirmed with the business data in the USA. Firms invest in modern renewable energy because larger sales more than compensate the decreasing unit costs which confirms the behavioural idea that innovators compounds qualities assumed attractive to customers in niche markets followed by decreasing sales prices due to the cost-reducing technical change as the innovations disseminate across markets.

Modern renewable energy evolved in 50 years, from a utopian perspective of a handful of innovators to the commercial mainstream, which provides an alternative for fossil fuels, and thereby enables the mitigation of climate change, if it continues to expand fast. For the fast mitigation of climate change with accessible energy larger investments in bioenergy, geothermal, marine, and hydropower are also needed. Rather than being picky in search of ideals, decision-makers in policies, businesses, and civil organisations can find ways to generate investments aligned to the societal demands.

Chapter 6
Diffusion of Renewable Energy

Does renewable energy evolve toward the global energy resource? This question is discussed using statistical data on the carbon content of energy resources, value-adding services, storage in energy networks, and the valorisation trends from 1990 to 2015. Decarbonisation of energy resources is negligible, and policies on hydrogen may not improve it but cause higher costs. A more promising trend is the growing value-added energy services based on renewable energy, called valorisation. Therefore, investments in energy storage and network stability are needed, based on the traditional infrastructure and modern one with batteries. If the valorisation of energy services as seen in the last 25 years continues during the next 25 years, along with the substitution of fossil fuels for renewable energy, energy services become accessible and CO_2 emission reduction enables mitigation of climate change.

6.1 Introduction

Renewable energy enables a reduction of the carbon content in energy consumption because these resources are low carbon. It also enables better energy services because widely attainable and some qualities are appreciated by consumers. During a period of time, these resources can generate less carbon than fossil fuels, called decarbonization, and the services can add value, called valorisation. A question is whether decarbonisation and valorisation can be observed as trends in energy consumption during decades in the past, thereby can be expected in the next decades on the global scale.

For an answer, the decarbonisation of energy resources and valorisation of energy services are assessed in 14 countries with more than 100 million citizens and globally during 1990–2015 as mentioned in the Introduction. Those 25 years are divided into the periods of low international prices of fossil fuels and emerging policies on renewable energy 1990–2004 and the period of high prices and supportive policies 2005–2015. The decarbonisation is measured as the decrease of CO_2

Y. Krozer, *Economics of Renewable Energy*,
https://doi.org/10.1007/978-3-030-90804-1_6

emissions and the valorisation of energy services is estimated as an increasing value of energy consumption in the monetary terms. They are assessed with data of the International Energy Agency (IEA) in 5-year intervals. The decarbonisation and valorisation are considered as trends when the changes evolved during a decade or longer. As the trends differ across countries, the differences between the countries can increase over time, it means diverging, or the differences can decrease, which means converging. It is assumed that the past trends in decarbonisation and valorisation continue in the future if the converging trends across countries are observed in the past, whereas diverging trends indicate high uncertainty about future trends. As mentioned in the introduction, the diverging and converging trends are based on the standard deviation of the growth rates across countries; a diverging trend is assumed when the standard deviation increases over time and the converging one is assumed when the standard deviation decreases.

After estimates of the decarbonisation trends, the costs of a few technologies that are promoted for the decarbonisation in the future are assessed. The valorisation is assessed based on the value added by the electricity services in the USA and EU and the valorisation trends across countries, which are followed by assessments of the distributed energy systems with storage. Finally, the impacts of valorisation on the income, energy and CO_2 emissions are estimated.

6.2 Decarbonisation Trends

The context of decarbonisation is that global CO_2 emissions grew annually by nearly 2% from 1990 to 2015, reaching about 32 billion tonnes in 2015 and 36 billion tonnes by 2020 as shown in Sect. 2.5. Regarding that growth, the questions is what are the trends, the drivers, and whether they diverged or converged across countries. For this assessment, three indicators for the decarbonisation are used on the global scale and across countries during the period 1990–2015. The first indicator is kg CO_2 emission per capita, called carbon consumption. Lower carbon consumption implies less environmental damages per person. This is influenced mainly by lower energy consumption due to the economic changes towards less energy-consuming sectors, energy saving within the sectors, and the composition of energy resources with fewer fossil fuels, but more renewable energy and nuclear power. The second indicator is the income per carbon consumption, measured in USD per kg CO_2 emission, called carbon performance. This indicator shows the generated income compared to the damages. A higher carbon performance, it is more income per carbon, means lower damage given income. Thirdly, kg CO_2 emissions per kWh energy consumption is called carbon efficiency. This one indicates the shifts of energy resources in consumption from high-carbon to low-carbon renewable energy and nuclear energy. The carbon efficiency compared to the carbon consumption indicates the importance of resource composition compared to other factors that influence the decarbonisation.

Those indicators are assessed. Table 6.1 shows the global and countries' CO_2 emissions in tonnes per capita, GDP-PPP in USD per kg CO_2 emission, and the

Table 6.1 CO_2 emissions in 2015 and annual average growth during 1990–2015 based on the IEA data

	Carbon consumption		Carbon performance		Carbon efficiency	
	Tonne CO_2/ person 2015	growth 1990–2015	USD-PPP/kg CO_2 2015	growth 1990–2015	kg CO_2/ kWh 2015	growth 1990–2015
World	4.4	0.5%	2.3	1.0%	0.20	0.0
USA	15.5	−0.8%	3.3	2.3%	0.20	−0.4%
Japan	9.0	0.3%	5.2	0.7%	0.23	0.5%
EU	6.3	−1.2%	5.6	2.6%	0.17	−0.8%
Russia	10.2	−1.4%	1.2	2.4%	0.18	−0.7%
Mexico	3.7	0.9%	2.7	0.5%	0.20	0.5%
Brazil	2.2	2.3%	5.2	−0.9%	0.13	0.6%
China	6.6	5.3%	1.0	3.7%	0.26	1.0%
Indonesia	1.7	3.4%	2.2	0.0%	0.17	1.4%
Philippines	1.0	2.2%	2.6	0.2%	0.17	1.7%
India	1.6	3.9%	1.2	0.8%	0.21	1.4%
Nigeria	0.4	0.8%	7.2	1.8%	0.04	0.4%
Pakistan	0.8	1.6%	1.5	0.1%	0.13	0.7%
Bangladesh	0.4	5.8%	2.2	−2.0%	0.16	3.0%
Ethiopia	0.1	3.3%	4.7	0.2%	0.02	3.1%

emissions per kWh. All those are shown for the situation in 2015 and the annual average growth during 1990–2015; the growth during low fuel prices 1990–2004 and high fuel prices 2005–2015 are also assessed and briefly addressed below but they are not shown in the table for the sake of readability.

Only a few countries determined the total global CO_2 emission as the USA, Japan, EU, Russia, China, and India covered more than 64% of it in 2015, but the diverging trends in total CO_2 emissions are observed which implies that this share declines. While the total emissions grew fast in China and India, they stabilised in the USA and Japan, and declined in Russia and the EU. Meanwhile, emissions grew fast in mid-income Brazil and Indonesia, and in low-income countries as Bangladesh and Ethiopia. Moreover, as total CO_2 emissions diverged across countries, high fuel prices enhance that global emission and the diverging trend because many mid- and low-income countries shifted from the traditional renewable energy to fossil fuels and within fossil fuels to the cheapest coal.

Global carbon consumption per capita increased in all countries, except the USA, EU, and Russia where it decreased, while all three had consumed more carbon per person than the global average in 2015. When growth rates of the carbon efficiency are subtracted from the carbon consumption per capita, the changes in energy saving can be separated from the changes in energy resources. The USA, EU, and Russia saved energy while other countries consumed more per capita. Particularly, China had high consumption per capita, on average more than twice as much carbon as nearly three times richer typical European. The carbon consumption per capita diverged across countries, whereas high fuel prices enhanced that, except in high-income countries and in Mexico.

The carbon performance varied across countries in 2015; for example, the Nigerian carbon performance was six times higher than the Russian one while both were large oil and gas producers and exporters. It grew in all countries, except Brazil, mainly due to shifts from energy-intensive industries to labour-intensive services. However, this growth of energy performance was lower than the growth of energy consumption in many countries, which implies the increasing damages of income. Meanwhile, differences in the carbon performances across countries are observed and they diverged. High fuel prices improved the carbon performance in most countries, but not in Brazil and Bangladesh.

High carbon efficiency of energy consumption means low CO_2 in kg per kWh. The carbon efficiency was low and improved in high-income while it was low and worsened in mid-income countries, whereas low-income countries had a higher carbon efficiency which also worsened. The carbon content per kWh declined in the USA, EU, and Russia, increased in all other countries, mainly because more coal and less traditional renewable energy were consumed. High carbon content is in the Chinese and Indian energy consumption. The fast growth of carbon content in energy consumption is observed in Bangladesh and Ethiopia as their economies shift to fossil fuels. The trends of CO_2 per energy consumption diverge across countries because the carbon efficiency improved in high-income countries while worsened in most other countries. High fuel prices had little effect on the carbon content in energy consumption, but several countries turned from natural gas and toward cheaper, carbon-intensive coal.

The global decarbonisation did not progress in total and per capita across countries. The carbon performance also hardly improved and the carbon content of energy consumption hardly changed globally. While the carbon consumption, performance and intensity improved in high-income countries, which were high consumers of carbon, they worsened in mid and low-income countries. Moreover, the trends across countries diverged and high fuel prices enhanced that mainly because mid-income and low-income countries increased consumption of coal compared to oil and gas. A leapfrogging toward renewable energy is not observed, though expected by some scholars (Levin & Thomas, 2016). The decarbonisation trends are not promising for the future despite political commitments and high prices of fossil fuels cannot enhance this trend as low as coal is cheap.

A question is if the coordinated policies can turn these trends toward decarbonisation across countries. Therefore, it is assessed whether similar trends are observed across the EU 28 countries as its carbon emission decreases annually and the member countries encompass high-income countries as Luxemburg and Denmark, as well as mid-income countries as Romania and Bulgaria. The EU defined policy in 2000 on CO_2 emission reduction by 30% compared to the 1990s in 2020; the EU-ETS is the main instrument for this enforcement. Meanwhile, the EU policy aims at 55% CO_2 emission reduction by 2030 and to fulfil decarbonisation in 2050. Given this policy, progress in decarbonisation and convergence across the member can be expected. Appendix 5 shows a similar table to Table 6.1 with the results for the EU countries based on the Eurostat statistics for the period 2008–2018; earlier data are incomplete. The highest carbon consumption per capita in Croatia is 4 times lower than the lowest one in Estonia, both mid-income countries, but there are also differences across high-income countries; for example, Sweden has low carbon intensity in the EU, whilst the neighbouring Denmark has a high one. The rate of reduction also varies; for example, Cyprus reduced its carbon consumption per capita by −3.8% and Spain by −3.6%, contrary to the increase in Greece by +2.7% and Luxemburg by +2.2%. The trends diverged across countries. Data on carbon performances are deficient, but show large differences across countries; for instance, 8 times higher performance in high-income Luxemburg than in mid-income Estonia. The carbon efficiency is similar across countries, but an unreliably low carbon per kWh in the United Kingdom. However, the growth varies from a +4.2% annual increase in Ireland – it means a decreasing carbon performance – to a −2.4% decrease in Romania. A diverging trend is observed. In effect, the diverging trends are observed within the EU despite its decarbonisation and policy coordination.

The decarbonisation of energy consumption is observed in a few high-income countries with high carbon consumption per capita, but hardly in mid-and low-income countries with lower per capita consumption. The trends across countries diverged. High fuel prices hardly contributed to the decarbonization while enhanced that divergence. The decarbonisation is difficult to steer across countries.

6.3 Hydrogen for the Decarbonisation

CO_2 emissions are abated through storage in plants but many policies focus on the technologies for carbon capture and storage (CCS), eventually with the reuse of stored CO_2 (CCUS), and conversions of fossil fuels into hydrogen with CCS. However, these technologies are in the pilot stage which implies that their energy efficiency, effects and costs are solely expectations. The CCS option is pursued despite higher costs compared to many decarbonisation options (Creyts et al., 2007).

The CO_2 abatement with plants refers to the cultivation of perennial plants because the growing plants enable net storage of carbon in their tissues and in the soil. While the growing perennial plants offer a potential sink of several tonnes of carbon per hectare, some studies assume up to 160 tonnes per hectare, the CO_2 sequestration depends on afforestation and input-output balances of the greenhouse gasses in forests. If the afforestation would be unconstrained on 4.9 billion hectares of farmland, the potential sequestration could be 29.4 billion tonnes CO_2 compared to 36 billion tonnes CO_2 emissions in 2020. Studies indicate that only 2.9 billion tonnes carbon equivalents is possible, assuming USD 10–150 per tonne carbon for the sequestration (Richard & Stokes, 2004) and about 4.9 billion tonnes carbon equivalents at USD 200 per tonne (Doelman et al., 2020). Ideas about mixed production of foods and forests emerge, but such agroforestry is rarely implemented, as yet. Thus, a large share of CO_2 emissions can be sequestrated through the afforestation for the standing forest but not all. Studies also suggest that forests capture CO_2 from the air but emit methane and other volatile organic compounds which are greenhouse gasses, thereby limits the positive impacts of sequestration (Popkin, 2019). Planting of perennial plants, including afforestation, is mainly constrained by investment costs but uncertain revenues. While investments usually exceed USD 1500 per hectare, incomes are generated only after several years of trees maturation (Reij & Winterbottom, 2015); bioresources from short-rotation coppices face competition from chemicals. An option is setting up the nature areas for leisure, tourism, education, and other income-generating activities, combined with CO_2 storage, biodiversity, and other common goods. Such carbon farming is assumed economic in the thinly populated areas, but the costs of nature areas expand in densely populated areas; for example, these costs increased from USD 20 per hectare to USD 1650 per hectare from 1995 to 2017 in the EU (Krozer et al., 2020). Afforestation with associated agroforestry, carbon farming, and nature areas can grow at about USD 100 per tonne sequestrated carbon equivalents.

A technological option is the capture of CO_2 from vent gas of large emission sources as coal-fired power plants, steel mills, and refineries, followed by storage underground. In theory, this CCS option can store nearly one-third of all emissions if enforced at all large emission sources but it needs additional energy and poses risks of toxic emissions during the storage and can cause earth movements thereafter. The additional costs are expected to be in the range of USD 15–90 per abated CO_2 tonne at coal power plants up to USD 30–270 per abated CO_2 tonne at gas power plants. However, pilot plants in Australia (Gorgon) and USA (Petra Nova and Kemper) suffer from deficiencies in operations which cause higher costs. The unit costs vary even more when it is aimed to reuse CO_2 because such CCUS operations depend on the sales of captured CO_2, which is uncertain (Metz et al., 2005); proposals for large scale reuse of CO_2 in horticulture, cultivation of algae, production of carbon fibres in plastics, building materials, and other valuable products and services are in infancy, so far (Climate Cleanup, 2020).

A decarbonisation option is application of hydrogen for replacement of fossil fuels and for storage of power from intermittent wind and solar energy. Given that this option is embraced by large-scale industries and policies in several high-income countries its costs per unit CO_2 emission reduction are discussed in more detail. The starting point is the conversion of methane in natural gas, called 'grey hydrogen'. By 2020, nearly all hydrogen is produced by this conversion. About 90 million tonnes of 'grey hydrogen' is applied for the production of nitrate and urea fertilizers. This conversion evolves in two steps. Firstly, methane (CH_4) with water (H_2O) is converted into carbon monoxide (CO) and hydrogen (H_2) in a process called steam reforming of methane (SRM). This is done at high temperature – up to 1100 °C – which consumes a lot of energy per hydrogen unit, Second, CO with H_2O water is converted into H_2 and CO_2 in a process called water to gas shift (WGS), which needs lower temperature and delivers sufficient energy for this conversion step. The SRM+WGS process together has a negative net energy balance; it is endotherm. Hydrogen, being three times more energy-dense compared to natural gas, can replace fossil fuels. It can also be used for energy storage; for example, provide backup for periods when insufficient electricity is generated from intermittent wind and solar energy based on the periods when excessive intermittent renewable energy is generated. SRM+WGS delivers four units of hydrogen along with one unit of CO_2 per methane unit along with large use of energy, water, catalysts, and other aiding compounds. This process generates CO_2 emissions and product hydrogen is explosive.

The alternative processes for hydrogen are in development. An option is the methane conversion with the carbon capture and storage (CCS) aiming to reduce CO_2 emission, called 'blue hydrogen'. Such SRM+WGS+CCS process is in the pilot stage. Whether this process can reduce greenhouse emissions is disputed. A life cycle assessment indicates a higher greenhouse impact of 'blue hydrogen' than natural gas because the impact of lower CO_2 is outweighed by higher methane emissions caused by losses in the production of natural gas (Howard & Jacobson, 2021); these global losses are huge as shown in Sect. 4.2. Another option is the methane pyrolysis, which does not involve oxygen (O_2) but this process needs much energy for high temperature and pressure in an oxygen-free condition. It avoids CO_2 emissions, but consumes much energy. A third alternative is the electrolysis of water, which generates hydrogen and oxygen with electricity in installations called electrolysers. This hydrogen production can avoid CO_2 emissions if electricity for the electrolysers is generated by renewable energy; labelled as 'green hydrogen'. Small-scale electrolysers are used in the fuel cells but progress is slow because these processes need high pressure from 30 to 70 bars, a large electric capacity from 40 to 80 kW per kg H_2, and large investments from USD 200 to 2000 per kW (IRENA hydrogen, 2020). By 2020, all global electrolysers in fuel cells remained lower than 1MW which delivered about 100 tonnes of hydrogen, though the electrolysers were applied for longer than one century.

Globally, about 756 MW electrolysers is under construction for hydrogen production by 2025 (Statista, electrolysers, 2021). About 150,000 MW is planned, but several plans are still negotiated, among other the largest hydrogen plant based on wind power in Kazakhstan (Seijlhouwer, 2021). The political and business ambitions are high but the planned capacity is an insignificant scale for the global consumption of fossil fuels. For illustration, a production estimate is made to illustrate the planned scale. Assuming 60 kW capacity of electrolysers per kg H_2 operating at 80% of capacity, the planned electrolysers could supply 17.5 million tonnes H_2. This is equivalent to 583 TWh energy based on 33 kWh per kg H_2. The planned capacity of 'green hydrogen' until 2030 can replace about 0.45% of 130 PWh fossil fuel consumption if everything goes on as it is planned; oxygen (O_2) is a byproduct. Therefore, it is a political illusion to assume that the production of 'green hydrogen' can be a major energy resource for industries before 2030 or even later. More realistic is application for the power storage of the intermittent renewable energy resources along with an accelerated electrification of industrial production. Moreover, this illusion has a price because the costs of electrolysis are high. It is costly. By 2020, the largest electrolyser of 100 kW in Bécancourt in Canada, which uses cheap hydropower for electricity, can deliver hydrogen at the price of USD 2.67 per kg H_2, including byproducts but excluding taxes. This is equivalent to USD 0.081 per kWh (FuellCellsWorks, 2021), which is about twice the hydropower price. A larger scale can reduce the unit costs, but the operating hours of intermittent wind and solar energy are lower than the continuous hydropower, which can cause higher unit costs of electrolysers. Therefore, nearly 10 dollars-cents per kWh can be expected for applications of hydrogen in industries and storage of wind and solar power. If it is pursued, permanent subsidies for large industries can be expected.

Given those technological alternatives for decarbonisation, a question is about what are their costs per kg CO_2 emission reduction, on the assumption that hydrogen can replace coal, oil, and gas. The costs of CO_2 emission reduction of hydrogen production are estimated with a simple input-output model of a hypothetical plant (Wikipedia, 2021); its inputs are heat, electricity, methane, and water, and outputs of hydrogen, water, oxygen, and CO_2. Appendix 6 shows the estimated inputs and outputs, energy balance, and costs per kg hydrogen. Several critical points should be mentioned. More inputs are used for operations of electrolysers than in the model which implies more residuals and emissions (Amran et al., 2017). Solely the production of hydrogen is considered whereas CO_2 emissions in the life cycle of the hydrogen production should be accounted for (Mehmeti et al., 2018). Pilots of hydrogen production with methane conversion and carbon capture show twice higher CO_2 emissions per kg H_2 than in theory (Rapier, 2020). So, energy consumption and CO_2 emissions in practice are higher than the ones estimated below. Furthermore, the estimated costs per emission reduction are higher because costly additional materials and energy are used (Harrison et al., 2010), and note that the estimates of the levelised costs of energy (LCOE) of various technologies for hydrogen production vary a lot (Sciencedirect, 2021). The costs of CO_2 emission reduction also depend on the applications of hydrogen; in particular, whether hydrogen is used for the replacement of coal, oil, or gas because their carbon contents differ. The estimated costs per kg CO_2 emission reduction are only approximations because data are imperfect.

As those issues in practices are neglected, the 'best cases' of hydrogen technologies are indicated in the Fig. 6.1. This graph shows the percentage CO_2 emission reduction and unit costs of four hydrogen production technologies: steam methane reforming with water to gas shaft (SMR+WGS) without and with carbon capture and

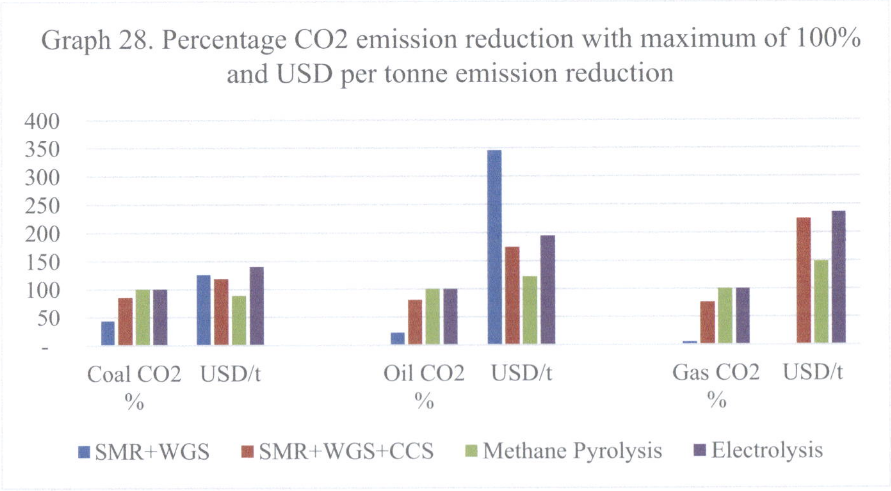

Fig. 6.1 Percentage CO_2 emission reduction and USD per tonne emission reduction

storage (SMR+WGS+CCS), Methane pyrolysis, and Electrolysis of water. Replacements of coal, oil, and gas are estimated based on their energy and carbon equivalents. The results are percentage CO_2 emission reduction compared to fossil fuels and USD per tonne CO_2 emission reduction; note that the unit costs of SMR+WGS for the replacement of gas are not shown in this graph as they are very high because reduce little CO_2. The electrolysis also delivers oxygen when water is split, but the income from sales of oxygen is neglected.

These estimates show that the SMR+WGS technologies reduce only some CO_2 emissions, and that the percentage emission reduction is higher for the coal replacement than for the oil and gas replacements because coal has a higher carbon content per mass. High percentage CO_2 emission reduction needs additional CCS but it consumes more energy and adds costs. The pyrolysis of methane and electrolysis of water reduce all CO_2 emissions, which means a 100% reduction as shown in the graph because do not involve oxygen as input. The unit costs of technologies with coal replacement are about USD 100 per tonne CO_2. The methane pyrolysis is presumably most cost-effective per tonne CO_2 emission reduction and the electrolysis of water is presumably most costly but the differences across those technologies are small compared to the replacement options. The replacement of coal is the most cost-effective option in USD per tonne reduced CO_2, given the highest carbon content of coal, and the replacement of gas most costly.

These estimates indicates that the lowest cost substitution of coal for hydrogen would cost USD_{2019} 80–130 per tonne CO_2 emission reduction, which could be expected in metal industries, coal-fired power plants, and other coal-based industry, whereas higher unit costs could be expected in the oil and gas industries. Even higher unit costs can be expected for the replacement of gasoline in vehicles or gas at households because additional costs in the distribution of hydrogen are involved. High distribution costs could be caused by the safety measures with regard to explosiveness of hydrogen. These costs could approach USD 1700 per tonne CO_2 emission reduction of gas, which is manifold higher than most other actions for the mitigation of climate change. This indicative assessment shows that high price of CO_2 through taxes, emission cap and trading, or subsidies are needed for hydrogen production with CO_2 emission reduction. The lowest unit costs – hydrogen pyrolysis with replacement of coal – are similar to the price increases of oil and gas during the 'shocks', but they are threefold of the CO_2 price in emission trading, or manifold of the average carbon tax in 2020. The lowest costs of hydrogen production are also similar to large-scale afforestation manifold costlier than many actions for higher energy efficiencies in industries and for the growth of traditional renewable energy that enables continuous heat and power through bioenergy and hydropower.

The costs of hydrogen production can decline through adaptations. For example, the European Commission expects that larger consumption of 'blue' hydrogen causes an additional USD 0.10 per kWh before sufficient decline of the costs of 'green' hydrogen (Cloete, 2020). The expectation that the unit costs of green hydrogen decline in the future can be verified with regard to the firms' interests in the improvements of hydrogen production. As mentioned in Sect. 4.3, the declining R&D costs per patent indicate firms' interests in future improvement, whereas the

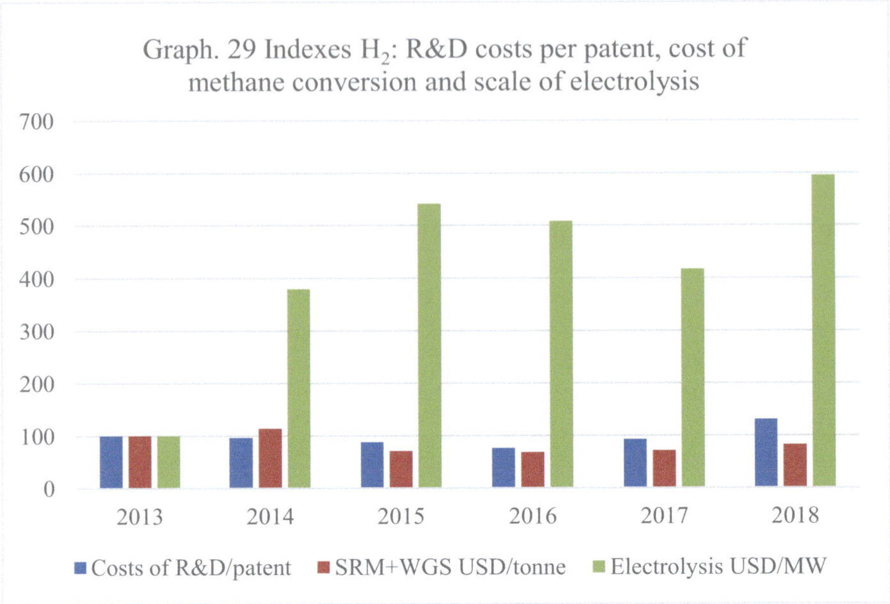

Fig. 6.2 Indexes H_2: R&D costs per patent, cost of methane conversion and scale of electrolysis

cost-reducing technical change in the past indicates possible changes in the future as the rate of change usually saturates after a while. Therefore, the R&D costs in hydrogen per patent in hydrogen and the cost-reducing change of SRM+WGS and Electrolysis are assessed for the period 2013–2018 with the IEA and OECD data. The results are indexed because the dimensions differ. Figure 6.2 shows the indexes.

While the R&D costs in hydrogen hardly changed, the number of patents somewhat decreased from 2013 to 2018, which indicates a declining business interest. The unit cost of hydrogen production with SRM+WGS technology declined only by -1% annual average during those years, which is negligible compared to double-digit cost-reducing technical change of wind and solar energy during that period. The scale of electrolysis capacity has somewhat increased, but the unit costs per installed capacity increased from 2013 to 2014 followed by fluctuations but did not decline. Apparently, there is not much business interest in the hydrogen production, neither high rate of the cost-reducing technical change in SRM+WGS nor in electrolysis of water as the scale grows. The sluggish progress does not come as a surprise regarding more than one century of failing adaptations of that technology. Scaling-up of electrolyser capacities from a few kWs in the past to several MWs in the future, presumably, generates a cost reduction because the capital-intensive installations are usually sensitive to the economies of scale, but radical cost-reduction should not be expected. Permanent subsidies are needed for hydrogen production while lower-cost options are not used. This production is instrumental for the industries that benefit from energy-dense resources and for the suppliers of natural gas that convert methane into hydrogen because complementary to the consumption of gas. However, all hydrogen options are costlier to small and medium size firms and households than directly available renewable energy for consumption. Meanwhile, hydrogen applications for the storage of electricity from intermittent renewable energy resources compete with rival options shown in the next section.

Technologies for the hydrogen production enable decarbonization albeit at higher social costs than many other options. Regarding the costly hydrogen production, low business interest, and low cost-reducing technical change large scale industries would need permanent subsidies paid by small and medium size firms and households, whilst the complementation of natural gas rather than substitutions for renewable energy is pursued.

6.4 Distributed Energy Systems

Massive consumption of wind and solar energy is obstructed by the intermittent power supplies and low energy density. This can be resolved with storage in hydrogen as discussed above, or by the balancing of electricity and heat with storage on-grid and off-grid in local networks with energy exchange. The latter option is referred to as distributed energy systems. Below, the costs of a few options for the distributed energy systems are assessed. The assessments focus on electricity because the costs of thermal exchange and storage for periods longer than a few weeks are usually higher than costs savings though technologies for thermal storage are available (Celsius City, 2020). In short, heat residue is usually too cheap for long storage. Nevertheless, there is progress in thermal energy storage and exchange in pilots reviewed in IEA-ETSAP and IRENA (2013) and popularised (CelsiusWiki).

Minor fluctuations in the electricity supplies to the grid can be balanced on the electricity network if the networks are large and robust; however, the intermittent supplies need additional electricity storage off-grid with information technology about power fluctuations and frequencies. As electricity consumption grew fast during the 1900s, technologies for energy storage off-grid became widely available. The traditional technologies are based on water storage when an excess of electricity is converted into water mass. Modern technologies use electricity for conversions in electro-chemical and electro-mechanical devices. The traditional hydro storage technologies are typically applied for the long cycles of storage, which means several weeks to years, and slow speed of loading, which continues during several hours to days. The mechanical and chemical storage is mostly used for short, repetitive cycles which are several hours to days, and fast loading from a few minutes to hours (WEC, 2020). These alternatives complement each other, but modern storage cannot replace the traditional one during the next decades because the hydro storage is manifold of modern one whilst manifold larger energy storage is needed for intermittent renewable energy.

Global storage capacity was about 0.192 TWh in 2020, estimated based on the database of the USA Department of Energy (DOE, 2021). Only one-thousandth percent of global electricity consumption was sufficient for the present continuous electricity production mainly based on the combustion of fossil fuels, nuclear power

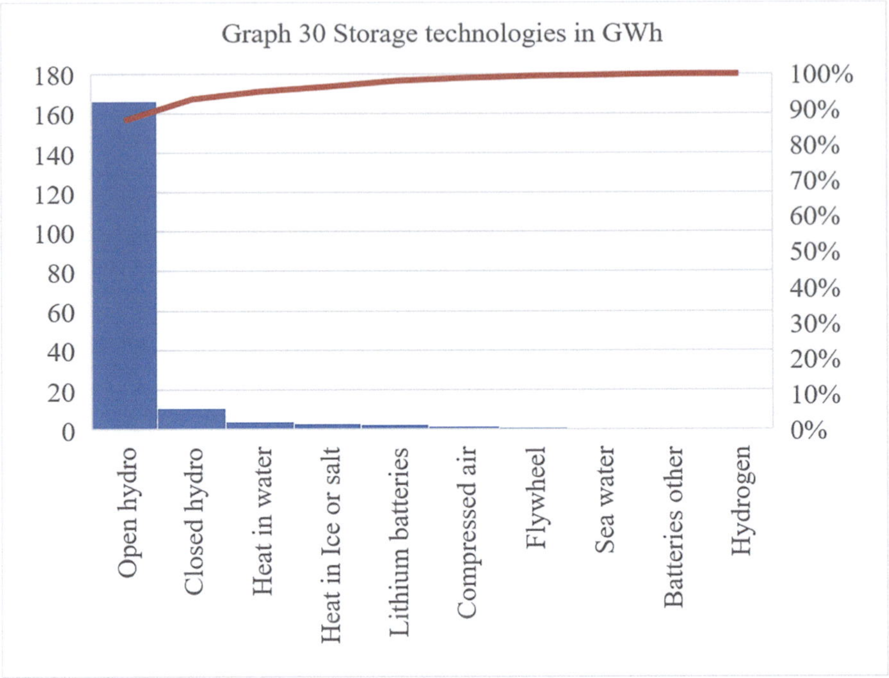

Fig. 6.3 Storage technologies in GWh

and hydropower. Minor fluctuations in supplies could be balanced on the grid. However, as more wind and solar energy were delivered to the grid from the late 1990s onward, more attention was given to electricity storage which invoked innovations in storage technologies. Figure 6.3 shows the global capacity of electricity storage by 2020 with main technologies.

By far the largest storage facilities are open water reservoirs at hydropower plants where the water level is pumped up during excess of power. During power shortage, the declining water level brings turbines in motion for electricity generation (87% of all power storage). Smaller facilities use a similar principle in closed water reservoirs in communities where water is pumped-up during an excess of electricity and released to flow on the turbines when more power is needed (6% of all power storage). Even smaller are storage facilities that use an excess of electricity for expansion of hot water and generate electricity when they are cooled down (2% of all power storage), or in the opposite direction by cooling or salination of water during the electricity excess (1.6% of all power storage). These hydro-based facilities for energy storage, mentioned in the ascending order of duration and speed of conversion, covered altogether more than 96% of global energy storage. The electrochemical devices covered nearly 2% of all energy storage. The most popular are lithium-ion batteries while zinc, nickel, vanadium, and other types of batteries are also used. Hydrogen-based electrolysers covers about 0.6% of all batteries which is 0.012% of the present storage. Hence, the hydrogen capacity must grow about

10,000 times for the present power storage and even more for larger intermittent renewable energy, which is unfeasible within a few decades. Compressed air, flywheels, and other electro-mechanical devices covered the remaining 2% of global storage. Detailed global inventory of the storage facilities encompasses a few dozen technologies while many are still in the pilot stage for testing.

For the future, a larger storage is needed for the growing electricity demands and diversified energy resources. Besides a larger scale of storage due to the demands, a larger share of intermittent energy resources in power generation and more frequent exchanges between producers and consumers are expected; for example, two-ways exchange for households with storage in car batteries, for local offices with storage in telework, and other distributed energy systems. It is estimated that global energy storage must be increased about twenty times to 2.5–4 TWh in 2030 for intermittent renewable energy, and larger and diversified power uses (US-DOE, 2020). Even larger storage enables downscaled technologies in networks of individual consumers in businesses and households; for example, the networks of PV on roofs, and batteries in electric cars. Storage in batteries attracted much policy support and investments as an alternative for inflexible hydro storage. For the envisioned storage capacity, the present storage in electrochemical and electromechanical devices must grow about 1000 times within a few decades, it is about 30% annual average, which is ambitious and possible. However, the costs increase as the hydro storage of USD 10–20 per kWh is about ten times cheaper than equivalent capacity in batteries (Roberts, 2019). An estimate for the USA is that modern storage at about USD 20 per kWh enables to accommodate all electricity production from intermittent renewable energy but this needs one-tenth of the costs of the late 2010s. It needs 20% cost-reducing technical change which is possible when the scale grows 30% a year (Ziegler et al., 2019).

The distributed energy systems with modern electricity storage are costly, but such energy services add the values of consumer autonomy, flexible use, consumption at the lowest tariffs, high status, as well as other functional and ethical qualities in energy consumption. Most batteries are used for such distributed energy systems on-grid and off-grid, sometimes also called 'behind the meter'. The scale of batteries on-grid used in the distribution of power is manifold larger than off-grid. The scale refers to the power density per volume and mass of the batteries; typical small-scale batteries are about 0.5 kWh per litre or 0.2 kWh per kg, whereas large scale ones go up to thousand times larger. For example, a consumer that generates 10 kWh solar power per day needs minimum 20 l of batteries, the equivalent of 5 car batteries, but in practice larger capacity in such off-grid, 'autonomous' system is reserved. An off-grid systems are presently a few times costlier than from the socket and more expensive than a diesel generator when power from the grid is not always available, but diesel generators have disadvantages in operations as maintenance, noise, and other pollution.

As in the case of hydrogen storage, it is assessed if the cost-reducing technical change in the modern electricity storage is observed and whether the sales of

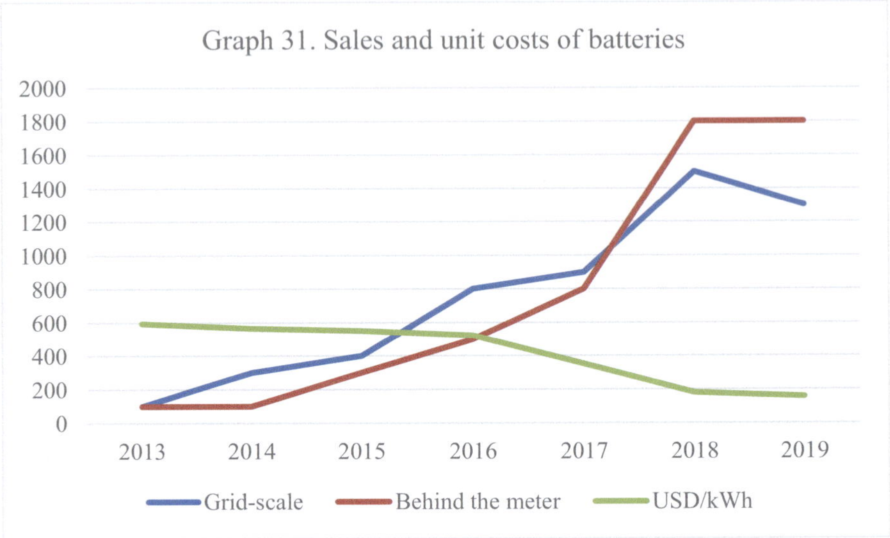

Fig. 6.4 Sales and unit costs of batteries

batteries on-grid and off-grid grew along with decreasing unit costs. The growth of batteries compared to the cost-reducing technical change indicates possible cost-reduction in the future. Figure 6.4 shows the indexed scales of batteries in GWh based on IEA data (IEA storage, 2021); and the unit costs of batteries in USD per kWh based on IEA data on the sales of Lithium-ion batteries (IEA-batteries, 2021). These indexes refer to the period from 2013 to 2017 (2013 = 100).

The global capacity of batteries increased 15 times from 0.2 GW in 2013 to 3.1 GW in 2019; while the batteries' capacity on-grid and off-grid grew by 1500% during those 7 years, all electricity based on renewable energy increased by 40%. Therefore, electricity consumption in the distributed energy systems grew several times faster than the renewable energy production, which underpins the value added by such energy services. The scale of the batteries storage grew by 60% annual average, whilst the cost-reducing technical change was nearly −18% a year and this rate was declining throughout that period; the unit costs of batteries declined from nearly USD 600 per kWh to about USD 160 per kWh, in the current dollar. The declining rate of change does not preclude ongoing cost-reduction. As in case of wind and solar energy, the saturation of the cost-reducing technical change is falsely forecasted. For example, the saturation of the battery costs is assumed to attain USD 280 per kWh in 2005 (Battery University, 2021), whilst the unit cost declined during subsequent years to nearly half that forecast. An encouraging indicator of the cost-reduction is the growth of R&D expenditures and even faster growth of patents in energy storage. Fast, cost-reducing technical change in the distributed energy systems can be expected if innovations in the electrochemical and electromechanical devices are generated and disseminated. For example, the unit costs decrease when the aircraft, shipping, and others businesses start with electricity consumption because such sales enhance technology development for energy-dense batteries.

Policies and investors are preoccupied with batteries and other modern storage technologies as their operations are fast, flexible, and convenient. This preoccupation is reflected in numerous policy studies on batteries, but not many on hydro storage, whereas the latter is decisive for storage capacity. For example, a comprehensive assessment of storage technologies is focused on the cost-reducing technical change in batteries, but hardly touches upon the performance of hydro storage (IRENA, 2017). That preoccupation invokes a rivalry between modern and traditional storage technologies for funding which seems to favour the modern one. However, unless traditional hydro storage is enhanced the growth of batteries alone is not sufficient scale in the next decades for the growing intermittent renewable energy. Moreover, the growth of intermittent power with power transfers between grid and storage can cause imbalances and more frequent failures on the grid in the near future.

Similar trade-offs are observed in the electricity networks. While the investments in the traditional networks stagnated, the investments in devices for the distributed energy systems grew, though such local networks cannot satisfy demands for electricity in the next decades. Figure 6.5 shows annual investments in the electricity networks which cover ICT and electric applications. These investments in ICT cover the 'EV chargers' which are devices charging of electric vehicles (EV), the 'smart grid infrastructure' which covers mainly automation of balancing on the grid, and the 'smart grid metering' which refers mainly to information and communication applications for consumers – these can be considered modern networks for the distributed

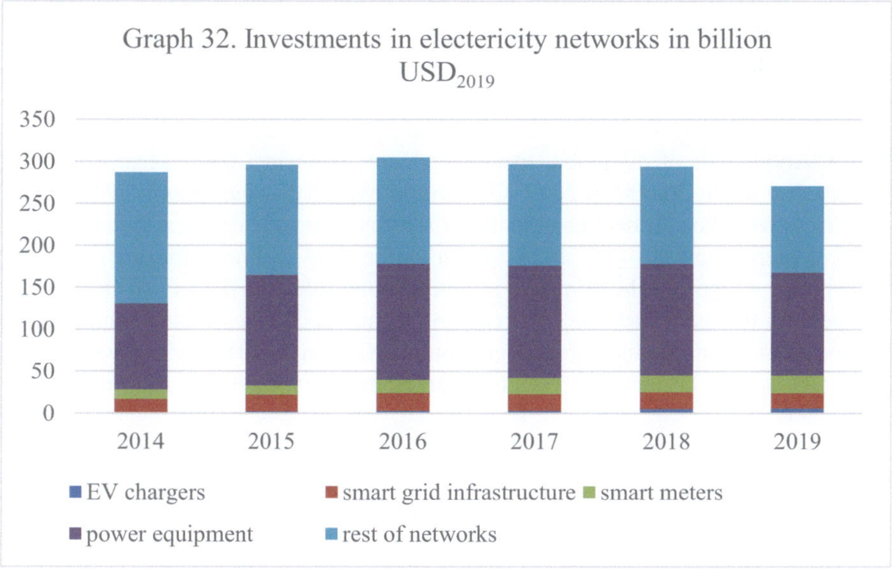

Fig. 6.5 Investments in electericity networks in billion USD_{2019}

energy systems. The investments in electricity networks also include the 'power equipment' with capacitors, transformers and suchlike, and 'the rest of the networks' with structures cables and so on – these refer to traditional networks. The IEA data is used for assessment of changes in these infrastructural investments (IEA smart grid, 2021).

While global demands for electricity grew, increasingly reached dispersed rural communities, and suppliers in intermittent solar and wind energy entered, global investments in the electricity infrastructures declined −1% annual average during 2014–2019. Within the average investments in all electricity networks of USD 290 billion per year, the investments in distributed energy systems grew about 10% annual average, reaching 17% of that average in 2019; the charging of electric vehicles grew particularly fast. Meanwhile, investments in the traditional electricity networks decreased by −3% annual average; in particular investments in the 'rest' category declined −8%, which means that masts, cables, and suchlike structural works are delayed. More frequent electricity failures on-grid and other malfunctions in the energy services should be expected in the next decades which provides incentives for the distributed energy systems, but it also obstructs the value addition by energy services when services cannot be delivered or perceived unreliable.

Within the investments in distributed energy networks, a growing share is allocated in the information and communication technologies for the metering, monitoring, planning, and suchlike tools in the energy demand management. For example, these expenditures in the USA grew by 17% annual averages from 2005 to 2017, starting at about nil and growing towards about USD 3.3 billion (Statista smart grid, 2021), whereas another study with a broader definition of the networks because including lighting and water heating estimated even 23% annual average growth (Pyper, 2017). The expenditures in electric mobility grew even faster. Electric mobility covers battery-based electric vehicles (BEV) and hydrogen-based vehicles with fuel cells (FC). After selling a few hundred BEV's called REVA, which were driving in India in the mid-2000s, the number of electric vehicles grew by 86% a year during 10 years from 20,000 in 2010 to nearly 4.8 million in 2019 (IEA EV, 2021). About half of all global electric vehicles were in China. Meanwhile, FC remained at about 20,000 in 2019. One would expect that the growth of BEV's is associated with the cost decline of batteries for this purpose; however, the correlation between them is low ($R^2 = 0.28$). A reason for the slow decline of the costs could be large policy support of the domestic production and sales of BEV's, which impedes competition that generates the cost-reducing technical change; those policies aimed to support the domestic industries (Reichmuth & Goldman, 2017). Moreover, BEV's hardly reduced CO_2 emissions because charged mainly with electricity generated by fossil fuels.

Energy services generated valuable distributed energy systems with batteries for electricity storage, information, and communication in energy applications and electric vehicles. Higher investments in traditional electricity storage and networks are necessary to overcome the stagnation in infrastructure for renewable energy and improve the performance of electricity networks.

6.5 Value-Added Energy Services

As mentioned in the section about energy performance and about benefits, electricity services deliver valuable qualities for lighting, sound, and other services whilst modern renewable energy can add qualities as backup for peaks in energy demands and consumption in remote areas, among other benefits for private interests. For the collective interests, the electricity services generated spin-offs due to fine mechanics, communication, and other income-generating activities whilst modern renewable energy reduced hazards of combustion, pollution, and CO_2 emission because a reasonable percentage of consumers are inclined to pay somewhat higher prices for renewable energy. As higher prices of services can be driven by the diversified qualities that are appreciated by customers, the growth of modern renewable energy can be comprehended as an innovation spur for higher prices due to higher qualities of energy services. Those qualities can be measured as the value addition by energy services. This value added is discussed largely based on an earlier paper (Krozer, 2019). The value addition is measured by sales minus purchases of energy and materials when the sales are equal to the sale price multiplied by the deliveries and purchases are the cost of energy resources multiplied by the volume of purchases. For example, the international price of natural gas in 2015 was below USD 0.02 per kWh, whilst the selling price to households in the EU was about USD 0.06 for gas and USD 0.14 for electricity without taxes, and USD 0.08 and USD 0.23 with taxes (Eurostat, 2021). This implies that the value added by gas services was USD (0.06–0.02) = 0.04 multiplied by 1146 TWh gas supplies, which is USD 46 billion. The value added by electricity services was USD (0.14–0.02) = 0.12 times 3458 TWh electricity supplies, which is USD 414 billion, or more when cheaper coal is considered.

Given the competition, the growing value addition by energy services during several subsequent years indicates improvements of the qualities in energy services; it is referred to as the valorisation of energy services. The valorisation of energy services and the role of modern renewable energy in it are estimated for the electrical services in the USA and EU during 2005–2015 years, which means during high prices of fossil fuels. While the energy services are estimated mainly based on the statistical data from the Energy Information Administration (EIA) for the USA (EIA total energy, 2021), the Eurostat data are used for the EU. The international resource prices are based on the EIA data for the USA (EIA prices, 2017); they are assumed applicable to the EU because such data is unavailable in the EU statistic, but the fuel mix is country-specific. The sales prices are averages of various types of households and businesses in the USA, as well as mid-size households, mid-size firms in the European Union including profit and tax. All prices are calculated in real USD_{2005} and the supplies to households and businesses in TWh. Table 6.2 shows the results.

The valorisation of energy services is observed in the USA and the EU households, but not in the countries' businesses. The costs of electricity to households were on average about 10 times higher than the purchase prices of energy resources in the USA and 31 times higher in the EU, both are including taxes; these costs were 6 and 16 times higher in businesses. While the costs of purchases decreased yearly on average −2.7% in the USA and −2.4% in the EU – this difference reflects their fuel mixes – the cost of services to households increased by a 1.7% annual average in

Table 6.2 Purchase prices, sale prices and production value of supplies to households and industries in the USA and EU during 2005–2015, accounted using the formulae below based on EIA and Eurostat data

USD_{2005}	Purchases USD/kWh		Households' sales price USD/kWh		Industries' sales price USD/kWh		Value added household USD billion		Value added industry USD billion	
Years	US	EU	US	EU	US	EU	US	EU	US	EU
2005	0.011	0.007	0.095	0.17	0.06	0.08	113	253	47	87
2006	0.010	0.007	0.101	0.17	0.06	0.09	123	269	50	96
2007	0.010	0.006	0.100	0.20	0.06	0.11	125	316	51	114
2008	0.013	0.008	0.104	0.21	0.06	0.12	126	347	52	124
2009	0.010	0.006	0.106	0.21	0.06	0.12	131	341	49	110
2010	0.010	0.006	0.105	0.20	0.06	0.11	137	341	50	106
2011	0.010	0.007	0.104	0.22	0.06	0.11	134	365	50	111
2012	0.008	0.007	0.104	0.21	0.06	0.11	131	351	49	101
2013	0.009	0.006	0.104	0.23	0.06	0.11	132	379	49	100
2014	0.010	0.006	0.105	0.23	0.06	0.10	135	370	50	96
2015	0.008	0.005	0.105	0.20	0.06	0.08	137	321	49	76
Annual average growth	−2.7%	−2.4%	1.1%	2.0%	0.1%	0.3%	2.0%	2.8%	0.5%	−0.7%

Value added $v_t = p_{s,\,t} \cdot q_s - p_{r,\,t} \cdot q_r$ for resources volume $q_{r,\,total} = q_{t,\,coal} + q_{t,\,oil} + q_{t,\,gas}$ and prices $p_{r,t2005} = p_{t,coal} \cdot {}^{q_{coal}}/_{q_{total}} + p_{t,oil} \cdot {}^{q_{oil}}/_{q_{total}} + p_{t,gas} \cdot {}^{q_{gas}}/_{q_{total}}$

the USA and 2.0% in the EU. Meanwhile, supply of electricity increased by 0.4% in the USA and 0.6% in the EU. In effect, the sales of electricity to households increased throughout that period in the USA and EU, whilst the sales of energy resources for electricity generation decreased. This implies that the value-added grew each year by 2.0% on annual average in the USA and 2.8% in the EU, which are equivalents of the additional USD_{2005} 2 billion in the USA and USD_{2005} 7 billion in the EU a year. In the industries, the value-added by energy services grew on annual average 0.5% in the USA whilst declined −0.7% in the EU. The electricity supplies decreased when cost increased, which resulted in yearly equivalents of USD_{2005} 0.2 billion in the USA and USD_{2005} −1.0 billion in the EU. The value added of energy services to households in the USA and the EU increased, not to businesses. Although studies suggest a sluggish behaviour of firms in energy savings when the energy prices increase in industries (Saygin et al., 2011), this assessment shows high sensitivity of industries to higher electricity prices.

Meanwhile, modern renewable energy grew faster than all deliveries of electricity. In the USA, solar energy delivery grew by 75% annual average up to 1.39 TWh and wind power by 44% up to 30 TWh in 2015; together they accounted for 2.3% of the electricity consumption in 2015. In the EU, solar energy delivery grew by 33% annual average up to 152 TWh and wind energy by 16% on average up to 302 TWh, which accounted for 27% of electricity consumption in 2015; inter alia, geothermal energy and biofuels are rarely used for electricity generation. The growing value added by energy services to household consumption in the USA and EU is also correlated to the growth of modern renewable energy between 2005 and 2015

($R^2 = 0.6$ for the USA and $R^2 = 0.5$ for the EU). Renewable energy is also more valued by the EU consumers regarding the increasing cross-country correlations between the electricity prices and the share of renewable energy in electricity consumption ($R^2 = -0.11$ in 2004 to +0.34 in 2015), possibly due to lower costs.

Other explanations of that trend are assessed. Higher prices of electricity services could be driven by market imperfections. An imperfection is behavioural slack. The growing income could reduce the sensitivity of households to higher energy prices, but the real personal income decreased in the USA and stabilised in the EU after the financial crisis in 2008. Subsidies could undermine competition, but the EU energy market was competitive regarding the high birth rate of energy firms and the decreasing share of the largest electricity producer; exceptions are Denmark, Germany, and Austria where local public firms merged, and the United Kingdom where private firms merged. Monopoly prices are possible in the United States as the scales of firms are larger and the birth rates of firms lower; yet, the sales prices grew slower than in the EU. The growing taxes on electricity could increase prices and undermine the competition, but the taxes on electricity did not increase in the EU during 2008–2015; these taxes rarely exceed 25% of the average consumer price and the sum of taxes and prices did not increase over time. The value-added without taxes grew on annual average faster than with taxes; they were 4.1% in households, 2.6% in business without taxes, respectively 2.8% and −0.7% with taxes. This implies that taxes did not trigger higher prices in the EU but this data is not found. It should be noted that the prices and taxes vary over time, per country and per scale of energy consumption; and they are regressive, meaning lower energy prices and taxes for a larger scale of energy consumption. The behavioural slack, poor competition, subsidies and taxes do not explain the valorisation of energy services in the USA and EU.

That valorisation of energy services generated incentives for innovations. The value-added by electricity services for consumption has increased yearly by USD_{2005} 2.4 billion in the USA and USD_{2005} 6.7 billion in the EU. For illustration, that additional income altogether is a few times higher than all expenditures on research, development, and demonstration projects for energy efficiency and renewable energy in these countries (OECD, 2021). The growing value-added implies that households paid for innovations by the electricity producers while businesses reduced purchases when prices increased. While the incumbent electricity firms were not major investors in the renewable energy R&D, the newcomers in the energy markets were the main innovators, as shown in Sects. 4.3 and 5.6.

The value-added of electricity services to households in the USA and EU grew, not to the businesses. That growing value-added was driven by the sales of wind and solar power which fostered the innovations spur in modern renewable energy.

6.6 Global Valorisation

The valorisation of energy services across countries is estimated during the period 1990–2015. Its context is income growth per capita and smaller income disparities across fourteen populous countries mentioned before, as estimated based on the

World Bank data as shown in Appendix 7. By 2015, a global citizen's purchasing power was about USD_{2005} 17,000, but a citizen of the USA had 35 times more purchasing power than an Ethiopian with USD_{2005} 1600 average. During that period, the purchasing power of a global citizen grew five times from 1990 to 2015. It increased twice in high-income countries compared to nearly 14 times higher in China, 5 times in India, and 4 times in Bangladesh and Ethiopia. These growth rates converged income across countries from 1990 to 2015. The convergence evolved fast during high fuel prices from 2005 to 2015, particularly after the financial crisis in 2008 that hit high-income countries rather than mid-income and low-income ones where the purchasing power increased; except Mexico that relied heavily on the USA capital market. The purchasing power grew threefold to fivefold faster than the electricity growth, which implies that more people across countries could afford higher value energy services. This is particularly important in the countries with low access to electricity; for example, only 26% of all people in Ethiopia, 60% in Nigeria, 63% in Bangladesh, 81% in India. Within this context of growing and converging purchasing power across countries, the valorisation of energy services as a global trend is underpinned.

The global valorisation of energy services is assessed based on the IEA data for those countries during 1990–2015 with the specifications for periods of unfavourable conditions for renewable energy from 1990 until 2005, and favourable conditions from 2005 to 2015 when the international prices of fossil fuels and policy support of renewable energy increased. Trends indicated by annual average growth rates during those periods were estimated. They were estimated for the energy performance measured in GDP-PPP in USD per kWh energy consumption, energy consumption in kWh per capita, and electricity consumption in kWh per capita. The valorisation of energy services is indicated by higher growth rates of the energy performance than energy consumption per capita and higher growth of electricity consumption over energy consumption. Besides the valorisation trends, it is also assessed whether these trends converged or diverged across those countries, which is indicated by the increasing or decreasing standard deviation of the valorisation growth across countries over time. Furthermore, it is estimated if high fuel prices enhanced the valorisation of energy services.

While the energy performance improved in all countries, except Brazil, it was faster than the growth of energy consumption per capita in 10 of those 14 countries throughout that period; it was slower in Brazil, Indonesia, India, and Bangladesh. During the favourable conditions for renewable energy, the energy performance was enhanced in 12 countries, except in in China and India. Meanwhile, the impact of high fuel prices on energy consumption per capita was mixed because decreased only in 7 countries. The valorisation of energy services is observed in most countries, which is enhance during high fuel prices. The energy performance and energy consumption per capita converged across the countries. Firstly, the purchasing power increased faster in mid- and low-income countries than in high-income. Secondly, high-income countries reduced energy consumption while mid-income and low-income countries increased it. These converging trends accelerated during periods of high fuel prices.

Meanwhile, the electricity consumption per capita was nearly 13,000 kWh in the USA, which was 16% of all energy consumption, but only 86 kWh, in Ethiopia, or 1% of all energy consumption. It grew in all countries, particularly fast in Ethiopia, Bangladesh, and Indonesia with a low share of electricity in energy consumption. Its growth was higher than energy consumption in 11 countries except for Brazil, China, and India. The electricity growth converged across countries, but high fuel prices impeded this growth and convergence. Those two indicators combined show the valorisation trend in most countries and the convergence across countries.

Less clear is the contribution of renewable energy to the valorisation of energy services mainly because of the trade-offs between traditional and modern renewable energy as mentioned in Sect. 2.4. While the share of renewable energy decrease in all 11 mid- and low-income countries throughout 1990–2015, mainly because the traditional renewable energy grew slower than fossil fuels, high-income countries enhanced modern renewable energy. Despite only 1.5% share of modern renewable energy in global energy consumption in 2015, the share grew by 5% annual average; during that period recall that this rate of fossil fuels substitution for renewable energy is sufficient for the mitigation of climate change if continues for several decades mentioned in Sect. 2.5. The share of renewable energy grew even faster than that in China, India, and the EU with a significant share of renewable energy in energy consumption in 2015, as well as in Brazil and Russia where the shares were close to nil that year. The growth of modern renewable energy converged across countries and high fuel prices enhanced that in several countries. However, the shares of renewable energy in energy consumption declined in several low-income countries. A global challenge is to reverse this trend by a slower decline of the traditional renewables along with faster growth of the modern ones.

If continuation of all past trends of fourteen most populous countries and the global one is assumed, income would grow, energy consumption would increase and become widely accessible, and CO_2 reduction would be attained if the growing renewable energy could substitute fossil fuels instead of the complementation. This is assessed by the extrapolation of the annual average changes in income, energy consumption, and renewable energy during 1990–2015 toward the period 2015–2040, on the assumption that fossil fuels are substituted for renewable energy. One extrapolation is without any change in the growth rates of renewable energy, labelled as the "continuing growth". Another one, labelled as the "decreasing growth", assumes that the growth rates of renewable energy are saturating. The saturation is calculated as a linear decline of the average growth rates during 1990–2015 down to 25% of that in 2040; this approximates a logistic function of a saturating market share over time. All calculations are done per country in 5 year intervals. Table 6.3 presents indices of the situation in 2040 compared to 2015 (2015 = 100); fossil fuels are not shown as they are related to renewable energy.

Both extrapolations demonstrate a 3.8 times larger global real income in 2040 than in 2015, and 2.0 times larger global energy consumption. This impact is driven by the valorisation of energy services which evolves in nearly all countries. However, energy consumption decreases in the high-income countries along with the slow growth of the purchasing power, whilst energy consumption grows in mid- and

Table 6.3 Extrapolations of annual average growth of income, energy consumption, modern renewable energy during 1990–2015 to 2015–2040 based on IEA data

	Income	Energy consumed	Continuing growth 1990–2015		Decreasing growth 1990–2015	
Compared to	USD–PPP per capita	kWh per capita	Renewable energy, Index of kWh	CO_2 percent of 2015	Renewable energy, Index of kWh	CO_2 percent of 2015
2015 = 100	2040	2040	2040	2040	2040	2040
World	378	201	10,263	46	990	79
USA	183	114	165	109	153	110
Japan	125	98	217	90	205	91
EU	145	97	540	23	348	53
Russia	118	82	102	89	94	90
Mexico	196	151	113	147	112	147
Brazil	196	213	424	71	225	205
China	938	349	50,884	–	4076	–
Indonesia	301	220	308	166	198	230
Philippines	250	156	149	153	124	206
India	427	254	1476	–	304	268
Nigeria	329	203	214	155	214	197
Pakistan	247	207	187	210	187	228
Bangladesh	327	267	135	296	135	335
Ethiopia	407	205	416	–	416	–

low-income countries along with the fast growth of the purchasing power. There is no difference between the extrapolations with respect to purchasing power and energy consumption per capita.

In the extrapolations to 2040, all renewable energy grows whilst modern one is the main contributor to the global renewable energy and CO_2 emission reduction because of its fast growth in several countries. As a result of that growth, China becomes the largest global renewable energy producer, followed by India and the EU. In the 'continuous growth', it is calculated that the global growth rates continue at the growth rates similar to those in the EU during the past 25 years; note that the EU growth rates are lower than the average growth rates in China, India, and Brazil during that time. It means that the continuation of the past growth rates into the future is plausible when government deliver what they promise. In effect, renewable energy would grow globally about 103 times, mainly modern renewable energy, whilst the global consumption of fossil fuels in 2040 would be reduced by 48% and the global CO_2 emissions would be reduced to 46%; both, compared to the 2015 level. China, India, the EU, Brazil, and Ethiopia would cause nil CO_2 emissions with much high energy consumption per capita, but lower in the EU. In the 'decreasing growth', renewable energy grows nearly 10 times, mainly modern one. The global growth rates decline below the EU average of the last 25 years. Global consumption of fossil fuels is reduced by 15% and global CO_2 emissions by 21%. This assessment indicates that the growth of income and energy consumption does not preclude mitigation of climate change if renewable energy is allowed to expand with the replacement of fossil fuels. Modern renewable energy is the driving force of changes.

No doubt that challenges emerge. Scarce space for modern renewable energy is a barrier in densely populated areas. For instance, the Chinese consumption of 121 PWh in 2040 can be covered with the available PV of 18% conversion efficiency and average irradiation of 1600 kW/m^2/year on 420,000 km^2. This requires 4% of the total Chinese surface, or 18% of its urbanised areas (Demographia, 2021). Although challenging, solar technologies can be integrated into buildings and infrastructure; meanwhile, the Chinese energy efficiency can improve twice to reach the EU. It is also argued that large-scale transport cannot be electrified but the experiences with scooters, cars, trucks, ships, and aeroplanes point to the contrary. Meanwhile, the energy density of electricity storage increases and unit costs decline fast which can continue for several decades. There is no shortage of innovators.

If renewable energy with the replacement of fossil fuels is pursued during the next few decades, at similar growth rates as during the last decades, incomes grow fast, energy consumption becomes widely accessible and emission reduction of CO_2 enables mitigation of climate change. These performances converge across countries.

6.7 Conclusion

It is assessed if the countries energy consumption evolved toward lower carbon content of energy resources measured by CO_2 emissions and higher value-added of energy services in USD-PPP, called trends in decarbonisation and valorisation when observed for a decade or longer. The assessment covers the years 1990–2015, divided into the period of unfavourable conditions for renewable energy 1990–2004 when fuel prices and policy support were low, and 2005–2015 when the fuel prices and policy support were high. Mainly the IEA data on 14 countries above 100 million citizens in 2015 and the world are used, the EU of 28 members considered as a unity. A converging trend across those countries indicates a continuous trend in the future, but a diverging trend suggests larger uncertainty about the global outcome.

Global decarbonisation of energy consumption is not observed. Global CO_2 emissions did not decline, neither carbon per capita nor carbon per unit of energy consumption. Sluggish global decarbonisation can be explained by the opposite trends across countries. Divergence in carbon consumption across countries is observed. While less carbon per energy unit is consumed in high-income countries, many mid-income and low-income countries turned away from traditional renewable energy to fossil fuels, which increased the carbon content of their energy consumption; the trends in carbon performance diverged. Furthermore, modern renewable energy competes for investments and policy support with traditional renewable energy in many countries. Moreover, energy consumption became more carbon-intensive during high oil prices from 2005 to 2015 because many countries shifted to cheaper, carbon-dense coal. As a result, decarbonisation does not progress globally, despite policies aiming at the reduction of CO_2 for the mitigation of climate change.

While the sequestration of CO_2 through afforestation hardly increased, more policy attention is given to the production of hydrogen for decarbonisation in industries. The production of hydrogen with methane in natural gas reduces some CO_2 in the product, but consumes much energy in the process and emits more methane in the life cycle. Carbon storage reduces more CO_2, but uses even more energy. Pyrolysis avoids CO_2, but this technology is not ready, yet. Electrolysis of water with electricity from renewable energy is considered as the solution, but this process needs much electricity and high investments per hydrogen unit. The scale of electrolysers is also insufficient for any significant decarbonisation in the next decades, even under the most optimistic assumptions. The perspective for cost-reducing technical change is also bleak regarding low R&D and declining patenting from the late-2000s onwards and slow cost reduction during the 2010s. Due to high energy density, hydrogen is useful for energy-intensive basic industries and energy storage delivered by intermittent renewable energy but is insufficient for decarbonisation and costly.

More promising changes refer to the distributed energy systems that can generate higher value in energy consumption. Larger consumption of intermittent wind and solar energy requires tenfold larger storage within several years. Therefore, investments are needed in large-scale, slow hydro storage by electricity generators and in more flexible storage in batteries and mechanical devices by consumers in businesses and households; long storage of heat is deficient and costly compared to low priced residual heat. Striking a balance between those two options requires policies. Fast, cost-reducing technical change in the flexible electricity storage is observed and perspectives for the future are positive given high R&D and many patents in modern renewable energy, storage, and power networks with ICT. Investments in such a 'smart grid' expand but in the traditional grid decline. Larger investments in the traditional grid are necessary to avoid failures in electricity supplies.

The valorisation of energy services is observed in the USA and EU, and a growing modern renewable energy contributed to it. This is a global trend because found in most countries and it is a converging trend across countries, which is enhanced by high fuel prices. The convergence indicates continuation in the future. This valorisation of energy services as a trend implies a higher energy performance mainly associated with the faster growth of electricity than fuels, and fast growth of modern renewable energy in electricity consumption, though several mid- and low-income countries lag behind. Extrapolation of the annual average growth in income, energy consumption, and resources from 1990–2015 to 2015–2040 along with the substitution of fossil fuels for renewable energy shows growth and convergence in income and energy across countries, larger energy access, and far-reaching CO_2 emission reduction.

The key factor is the replacement of fossil fuels by renewable energy. The challenge is to resist pressures of the interests in fossil fuels through the elimination of protection with subsidies for fossil fuels and enhancement of innovations in renewable energy that can cope with scarcity of space and minerals, wasteful consumption 'not-in-my-backyard' syndrome, and other challenges.

Chapter 7
Toward a Fair, Clean Energy

Mechanisms of change toward renewable energy in production and consumption are reviewed in this book; meaning, the socio-economic processes that generate innovations in energy resources and services. Such mechanisms of change enable policies in line with societal interests. This review refers to assessments of changes in the past, mainly based on the interpretations of statistical data starting from the early 1800s and ending in the late 2010s. Several mechanisms of change are pinpointed without prioritising preferences or hierarchies for actions, as they collectively represent various situations across countries, interests, and social groups. Diversity and value of renewable energy services increases. The main message is optimistic. The continuation of past trends, driven by those mechanisms, would enable the generation of renewable energy that is hundred times larger, earning an income per person of nearly fourfold and doubling energy consumption per person in half the CO_2. This perspective is attainable if people uphold preferences for value-added energy services and authorities resist protectionism in energy production.

7.1 Context

A question is discussed whether income growth, accessible energy, and far-reaching emission reduction of greenhouse gases are possible within one generation with the help of renewable energy. Therefore, mechanisms of change in the sense of methods that drive socio-economic changes are assessed with a focus on the mechanisms that generate or impede shifts from fossil fuels based on coal, oil, gas, and nuclear resources to renewable energy based on traditional uses of biomass, waste, and hydropower, as well as modern renewable energy using geothermal, wind, solar, and biofuels. As renewable energy is considered from the perspective of innovations that emerged in the past, the statistical data on these innovations is interpreted rather than a modelled future; the future is fundamentally uncertain. The interpretations are based on mainstream, neoclassical argumentation pre-occupied with the market prices,

Y. Krozer, *Economics of Renewable Energy*,
https://doi.org/10.1007/978-3-030-90804-1_7

evolutionary theorising about institutional decisions, and behavioural thinking that is focused on private and social values. Authoritative databases on energy technologies, production and consumption are used. Robust changes on the global scale, and across countries over a decade or longer are considered. Particular attention is given to the countries with more than a hundred million inhabitants, EU28 is considered as one country. These are high-income USA, Japan, and the EU; mid-income Russia, Mexico, Brazil, China, Indonesia, and Philippines, as well as low-income India, Nigeria, Pakistan, Bangladesh, and Ethiopia.

7.2 Available Resources and Lower Pollution

A prime issue is whether there are sufficient energy resources available for the growing energy consumption, and if sufficient pollution reduction with help of renewable energy is attainable. The available resources in renewable energy based on biomass, hydro, geothermal, wind, and solar energy are sufficient for annual energy consumption that is more than 2000 times larger than the one in 2015 with available technologies, taking into consideration availability of energy on land. While those renewable energy resources are available all over the Earth, facilitating downscaling of technologies for local uses, their energy density is low, they cannot be dispatched instantly and supplies are intermittent as they vary over time. In general, renewable energy resources are inconvenient for large-scale, energy-intensive production but for dispersed applications. Plants can convert these resources into energy-denser biomass; which can be processed into even energy denser and cleaner biofuels. However, the global stock of plants has been declining for centuries, as more forests are felled to allocate space for agriculture, husbandry, transport, housing, and other activities that undermined bioresources by volume and diversity. Fossil fuel resources based on coal with peat, oil, gas, and nuclear resources are substantially energy-denser than renewable energy, due to millions of years fossilisation of plants and volcanic eruptions but their availability is limited to particular spots on Earth where they are mined. They are scarce as their availability for future energy consumption is also limited. Present energy consumption cannot be maintained for a period longer than a few centuries, even with optimistic assessments. The exception is coal, whose stock is sufficient for a longer period.

As global energy consumption grew, more biomass, hydropower, wind, animal and human power, and other renewable energy were consumed along with larger consumption of coal, oil, and gas which became the largest energy resource by the early 1900s. Those resources are followed by nuclear power and modern renewable energy based on geothermal, wind, solar energy, and biofuels. Contrary to the mainstream assumption about the substitutions for cheaper resources and better technologies, the energy resources complemented each other in energy consumption. As a result, nearly all resources in renewable energy and fossil fuels increased throughout the last centuries. As they grew at various growth rates the resource composition of energy consumption changed while nearly all energy resources were maintained, not only globally but also within one country. Oil and gas grew next to biomass and coal during the late 1800s, but two latter lost the market share within 50 years, which implies that fast technological shifts after an incubation period are possible. This complementation mechanism can be explained by impediments posed by suppliers and by inventive entrepreneurs searching new markets. Given that innovations depend on the vested suppliers that have no interest in novelties, unless they share in profits, there exists a resistance to changes. Furthermore, the incumbent interests find specialisations through innovations in market niches, which diversifies energy resources.

Driven by consumers know-how about applications of energy resources in various businesses and households, those backward and forward linkages generated energy services. That know-how enabled the diversification of energy resources along with specialisation; for example, coal use in power plants and oil in mobility. Along with the expansion of those fossil fuels, new fossil fuels as gas for households and nuclear energy for electricity emerged. Furthermore, traditional renewable energy grew based on hydropower and modern renewable energy emerged and expanded during the late 1990s. Electricity grew faster than income and energy consumption during the 1900s, increasingly based on modern renewable energy from the 1990s onwards. Given the growth of knowledge-intensive services in energy, economic assessments should refocus from the energy resources towards services. The present policy focus on the energy-intensive industrial processes should shift towards value addition by energy services generated in consumer networks. The value-added by energy services with renewable energy increases welfare despite higher costs of energy resources.

Given the complementation mechanism in energy resources while rare substitution, CO_2 emissions caused by fossil fuels increased throughout the last centuries. Saturation of the growth of CO_2 emissions was observed in the late 1900s due to the consumption of natural gas and renewable energy. Although the required emission reduction for mitigation of climate change is disputable and data in the authoritative databases vary, far-reaching emission reduction is needed. About 75% emission reduction in 2050 is needed to attain the level of global CO_2 emission before the 1990s; it is when CO_2 in the atmosphere fluctuated within range. This is possible at 5–6% average annual substitution of fossil fuels for renewable energy during the next 30 years. Therefore, perseverance in policies aiming at the mitigation of climate change is more important than the introduction of a particular technological panacea, because diverse technological alternatives for this purpose are available.

A major challenge faced by the international policies on CO_2 emission reduction is the diverging composition of energy resources between high-income countries where energy consumption based on fossil fuels has saturated, and mid-income and low-income countries where that energy consumption grows. From the 1990s on, fossil fuels grew fast in low-income countries that consumed much traditional renewable energy. Meanwhile, fossil fuels and modern renewable energy grew fast in mid-income countries. In high-income countries, mainly modern renewable energy grew. The share of high-income countries in the global CO_2 emissions declined whilst the share of mid-income and low-income countries increased. Presently, global CO_2 emission largely depends on the six largest CO_2 polluters – USA, Japan, EU, Russia, China, and India in ascending order of income per person. As those diverging trends impede international coordination these six countries have a particular responsibility for turning the growth of fossil fuels into a substitution for modern renewable energy which enables far reaching CO_2 emission reduction.

7.3 Increasing Energy Performance

When industries expanded in Western Europe in the early 1800s, higher incomes enabled a larger consumption of food, clothes, energy, clean water, sanitation, and other private and public goods; which reduced mortality. While the income grew at various rates across countries, energy consumption diverged even more as some countries developed energy-intensive economies, whilst others remained energy-extensive. By the early 1900s, knowledge-intensive services were the main drivers of income growth. As service grew fast in the early 1900s in North America and Europe, their energy consumption and CO_2 emission were increasingly decoupled from income; followed by several Latin American countries and Oceania in the mid-1900s and finally by Russia, Africa and Asia in the late-1900s. New energy services expanded in the 1900s as consumers in businesses and households used more automotive engines and electrical appliances, as well as specialised suppliers delivered electricity, refined oil for fuels and chemicals. The drivers of growth and diversification of energy resources and energy services are explained from the mainstream, evolutionary, and behavioural perspectives.

From the mainstream perspective, scarcities of energy resources would cause higher prices followed by shifts to alternative resources, but experiences during the last two centuries indicate that policy interventions are more important than scarcities and market prices. Oil production grew fast during the declining prices in the late 1800s, steadily during the stable prices until 1970s and during high prices and low prices thereafter. Policy interventions in favour of oil are observed in the mid-1800s when oil was produced as a replacement of wood and agri-residues; for example, support to oil by taxes on rival alcohols and the legalisation of oil and gas monopolies. High oil and gas prices during the 'oil shock' decade between 1974 and 1984 when modern renewable energy was introduced, were driven by price agreements of oil producers. High oil and gas prices during the 'demand shock' decade between 2005 and 2015 when modern renewable energy was commercialised, were driven by the policy of economic expansion in Asia rather than oil and gas scarcity. Higher prices of the incumbents enabled the introduction of modern renewable energy resources, but their growth was a result of social engagement state interventions; for example, emerging energy cooperatives subsidies for modern renewable energy in several countries. Usually, the high prices of energy resources invoked novelties, yet entrepreneurial initiatives with policies generated their growth.

The evolutionary argumentation underlines the engineering capabilities in converting of energy resources into energy-denser products and energy-efficient services, while these capabilities are fostered by policies. The effect-increasing and cost-reducing technical changes as results of those capabilities are pinpointed. These capabilities increased energy efficiency and reduced costs of conversions per energy unit which, in turn, generated welfare and enabled larger energy consumption. This rebound effect caused that fast annual cost-reducing technical change in energy production was faded to minor global changes in total energy consumption and the carbon content of consumption. This string of thinking can explain the dissemination of initially costly innovations, not the reason why purchasers buy initially costly innovations whilst cheaper ones are expected.

An additional explanation of innovations refers to the value-added by energy services. While novel services are often costlier in purchases, they deliver qualities that reduce the cost during consumption; for example, provide larger power and suchlike functionalities, as well as autonomy in energy consumption and other ethical attributes. In most countries, the value addition in energy services grew from the early 1900s on, as shown by the increasing energy performance, measured in US dollars per kWh that is reciprocal to energy intensity. Meanwhile, the carbon performance – indicating environmental damages – diverged across countries. High-income countries with high carbon per capita reduced carbon in energy resources, whilst many mid-income and low-income countries increased that because turned away from biomass to fossil fuels. While the energy performance increased in most countries throughout the 1900s, it decreased in nearly all large producers of fossil fuels. These observations prove that costlier energy resources can generate beneficial energy services and that the vested interests in the production of fossil fuels effectively obstruct such changes. Turning focus in the economic policies from the industries in energy resources toward value adding energy services contributes to welfare and emission reduction.

7.4 Higher Chances for Innovations

Possibilities to pursue innovations in renewable energy are assessed with respect to the lead-time in development based on brief historical review of basic energy technologies. The lead-time of successful energy technologies is indicated, starting from the novel energy products (inventions), through novel businesses (innovations) and ending with sales of the novelties in a few countries (diffusion). Between 1800 and 2000, novelties in bioenergy, coal, electricity, mineral oil, natural gas, hydro, nuclear, geothermal, wind and solar power are reviewed as 29 basic energy technologies. Among them, 19 matured during 6–9 decades, a typical lead-time for maturation. This refers to the following technologies: biodiesel in bioenergy, large steam engines, coal-based boilers, fluidised bed combustion in coal, batteries-direct current, rechargeable batteries, alternating current in electricity, oil wells, four-stroke engines for gasoline, kerosene for light and aviation in oil, gas heating, thermal reforming in gas, hydropower turbines in hydropower, geothermal power plants, heat pumps in geothermal energy, horizontal and vertical wind turbines in wind energy, solar boilers, and concentrated solar power in solar energy. The time span from inventions to innovation, as well as from innovation to diffusion were 3–5 decades each and usually included several new designs and applications. Four basic technologies matured within 6 decades. All these were enhanced by military interests. These include a shift from coal to diesel oil by the British Navy in the 1910s;

nuclear plants whose high energy density was useful for ships and submarines of the USA and the Soviet Union; lasers in electricity pursued by the USA as a shield against missiles; and the photovoltaics in solar energy used for the space shuttles of the USA and the Soviet Union. The lead times of six basic energy technologies were 10 or more decades. These were: industrial bio-alcohol and biogas in bioenergy; hydrogen with fuel cells in electricity; tar sand in oil; and gas pipelines and fracking of gas in natural gas. The long lead-time was mainly caused by the societal concerns about explosions and pollution; for example, explosions in the industrial distillation of alcohols in the early 1800s and in gas winning during the 1800s, soil pollution caused by the fracking during the 1900s. When politics dismissed such societal concerns, diffusion stagnated after accidents as is the case with nuclear energy after two decades of booming sales. Long lead times imply that the mitigation of climate change cannot be based on new technologies but must use available renewable energy and that the societal concerns about safety and environment are serious considerations in the development of technology in order to prevent stagnation in diffusion.

Given the long maturation time, opportunities for innovations in a short term are assessed based on the R&D expenditures and patents in energy, globally, and in the selected countries mentioned above from the 1970s to 2019. Most R&D is spent by high-income countries. Meanwhile, China and India catch up. The R&D expenditures on energy resources were in line with their share in consumption, except for the R&D of nuclear energy that was many times higher than its consumption share. Expenditures on fossil fuels, energy efficiency and renewable energy are also sensitive to the market prices as they have fluctuated with the fuel prices. The R&D on nuclear power declined, on hydrogen increased followed by a decline, from the mid-2000s on and on the distributed energy systems steadily increased. Patents express the R&D results, thereby indicate tangible business interests. The number of patents is more or less in line with the R&D expenditures across those R&D areas but it is lower in nuclear power and higher in renewable energy, and the distributed energy systems. The R&D per patent were similar across countries, while Philippines and India R&D were more efficient than in high-income countries. Most business interest is expected in the distributed systems based on PV. Meanwhile, the business interests in development of fossil fuels are low, and the interests in nuclear energy and in hydrogen decline. Efficient R&D in several mid-income countries is a source of innovations in the near future.

Given those innovative opportunities and business interests, possibilities for selecting promising R&D projects in the period of 2007–2016 were assessed. The assessment addresses the chance that the particular R&D project obtains a patent, that the patent generates a start-up and that the start-up survives market competition during that period of ten years. This is estimated for all R&D, for energy R&D and for renewable energy R&D, insofar data are available. Experiences in the largest EU subsidy program for R&D – Horizon 2020 show that 14% of all project proposals are approved for subsidies, but only 1% of them obtained a patent, which is similar to the statistical data on all R&D projects and patents in the EU. In other countries, the chance to obtain a patent per R&D expenditure was higher, and more starting firms used patents; however, few of these start-ups survived over ten years. As a result of high survival rate, the EU innovators performed ten times better per million USD on R&D in all sectors than in the USA, Japan, Russia, and China, the EU performance was nearly as good as the Korean. Yet, the chances that an R&D project costing a few million USD would generate a surviving innovator was close to nil in all countries. Similar is observed for the energy R&D projects. Within the EU, the chances that an R&D project delivers an innovation is indicated by multiplication of the ratio's patents per R&D, start-ups per patent and survivors per start-up in all sectors and in the energy sector that covers mainly business opportunities in renewable energy and distributed energy systems. The chances in all sectors are close to nil because −1% to +1% across the EU countries. When innovations are demanded in societies the chances that R&D project generate successful innovations increase to a few percent, or higher, as in the cases of renewable energy and distributed energy systems. Therefore, assessments of R&D proposals are shots in the dark from the innovations perspective and policies should turn toward more successful methods. Societal demands for innovations generate substantially higher chances of success and should be a guiding principle in the R&D policy.

An innovator must spend much money and time on R&D and market introduction of its novelty before generating incomes. For this purpose, it needs private and public investors in addition to own funds and collaterals for loans. Such investments are risky because many novelties fail in generating sufficient income. The private investors cover only a few percent of all risk-taking capital. By far, the largest external funds for innovations are obtained from public investors through subsidies. A subsidy provides upfront funding. Therefore, public investors share the innovators' costs and risks but do not share in the income of the successful innovators. While the innovators are subsidised if their novelties are assumed attractive for the common good after detailed assessments, the promises cannot be verified before consumption. Regarding the low chance of success, the subsidising is gambling with public money. An alternative is setting boundary conditions for innovations based on the societal demands instead of the assessments with a high price guarantee for purchasing of the novelties that meet these conditions. Such guarantees for the purchases of renewable energy were highly successful and enabled broad public participation. Far-reaching demands for renewable energy with the guarantees for innovations in renewable energy increase the chances that R&D projects turn into successful innovative businesses.

7.5 Improvements in Decision-Making

The high fuel prices during 'oil shock' and 'demand shock' hardly influenced energy consumption that continued to grow, neither did the substitution of costly oil and gas for cheaper coal and nuclear power but they did trigger modern renewable energy.

However, high fuel prices were not decisive for the growth as modern renewable energy grew during low prices of the 1990s, followed by the faster growth during high prices after 2004, while less costly hydropower and bioenergy and waste grew slower throughout the period of low and high fuel prices. Within the modern one, well-known geothermal energy grew slower than costly wind energy, solar energy, and biofuels. The high prices of fossil fuels enabled the introduction of modern renewable energy, but other factors generated their fast growth during more than twentyfive years of low and high fuel prices from 1990 to 2015.

The policy support is often pinpointed, but this support is inconsistent because mainly fossil fuels obtained subsidies on-budget and tax relieves off-budget, which were many times larger than all support to renewable energy. The subsidies were delivered during periods of low and high prices of fuels without any tangible impact on low-income in mid and low-income countries, neither on energy consumption, but they effectively obstructed the growth of renewable energy in the countries with high subsidies for fossil fuels. Meanwhile, the global CO_2 price was close to nil as policies rarely taxed CO_2 emissions from fossil fuels whilst that price increases during those 'shocks' were equivalent of about USD 90 per tonne CO_2; the taxes only in two countries approximated this market prices. Furthermore, the average price of the CO_2 tax was only a few USD per tonne and the CO_2 price of emission trading was about a third of that equivalent. A tax would need to be high and combined with an emission cap in order to be effective. Moreover, high taxes and a cap with emission trading can invoke 'fast and dirty' actions as carbon storage along with additional consumption of fossil fuels instead of renewable energy and energy efficiency. Such actions contribute to the mitigation of climate change because store CO_2, but impede renewable energy and obstruct welfare growth because they do not add value but cause higher social costs. Therefore, erasing the support to fossil fuels combined with income compensation to the poor and enhancing renewable energy is socially fair, environmentally the most effective and economically, the cheapest energy policy for accessible energy with the mitigation of climate change.

The policy support for renewable energy is also ambiguous as it increased during times of high fuel prices and declined during low prices when it was most needed. The 'oil shock' invoked policy support for private investments in bio-alcohol production from wood, cane, corn, and other crops, but when the support declined during low fuel prices, these investments faded away. During the 'demand shock' period, modern renewable energy was supported with price guarantees for deliveries of renewable energy to the grid, and subsidies during the 2010s; both are usually called 'feed-in tariffs'. They fluctuated in most countries and declined when fuel prices decreased, which is inconsistent with the objectives of such policy support. Several countries with supportive policies prior to that price increase during the 'demand shock' gained as they became global leaders in the technologies for modern renewable energy during the 2010s. Public investments also generated renewable energy and infrastructure but their scale is disputed. High price guarantees for the delivery of renewable energy during low prices of fossil fuels would foster consistency of policy.

Modern renewable energy grew during low fuel prices mainly because local stakeholders had interests in expanding new economic activities in energy. While the motives of innovators and scales of their operations varied, most initiatives were ethically motivated and involved various stakeholders not per se seeking for the maximum profits. This way, local interests in support of modern renewable energy are generated, which often turned into firms contrary to the idea of scaling up of incumbent firms for larger energy efficiency. High-value energy services with public participation were successfully implemented in the EU. Due to the initiatives with policies for price guarantees, a few thousand additional renewable energy companies were generated annually in the EU during the economic crisis after the financial collapse, in 2008; meanwhile, a negligible addition is observed in the USA despite as large subsidies. These start-up companies in the EU delivered large employment in value-added energy services. They are supported at low social costs measured by the costs per job; thereby, affordable in all countries.

While from 1990 to 2015, the consumption of modern renewable energy grew globally by a few hundred times, development patterns changed across countries. The USA, Japan, Philippines, and other countries pursued the market-based system based on the corporate decisions which expanded fast in the early 1990s, and slowed down when profitability was disappointing during times of low fuel prices. Meanwhile, the policy-based system emerged and grew fast in China and Indonesia among other countries, where authorities determined investments. The stakeholders-based system of decision making, wherein the investments were driven by collaboration between private firms 'public institutions' and communities emerged in the EU and India among other countries. The latter deem successful measured by the global market share and stable with respect to the fluctuations in fuel prices and changing policy priorities, though it is not the fastest in decision making. The growth of modern renewable energy is motivated by the ethical considerations about social responsibilities as well as tangible benefits revealed in socio-economic behaviour. The main benefits are linked to a widespread availability that enables the distributed energy systems and opportunities for the generation of local business and employment. Larger renewable energy also reduces the health impacts of pollution and mitigates climate change.

Strikingly, the consumption of costly renewable energy based on wind and solar energy grew faster from the 1990s on than the cheaper one based on hydropower, biomass, and geothermal energy. This cannot be explained by policy support as it was not larger for wind than for other renewable energy resources, though the support for PV was higher in a few countries. The fast growth of wind and solar energy can be explained by much larger investment in the production capacities, which generate better capacity utilisation and cheaper equipment, thereby a fast effect-increasing and cost-reducing technical change. The fast growth of modern renewable energy is encouraging for the mitigation of climate change. However, the slow growth of bioenergy, waste, and hydropower is worrisome because these two constitute nearly 90% of all present renewable energy; thereby their growth is decisive for the mitigation of climate change within the next generation. All available technologies in renewable energy are necessary to attain far-reaching demands; hence, policies should not be picky but encourage designs that prevent negative impacts. Although it was often projected that the cost-reducing technical change of the wind and solar energy technologies would decline, that change remained as high as a two-digit rate for a few decades, better designs of application with hydropower, bioenergy and marine technologies can also progress. Apparently, the progress in technologies can sustain for a long time if high investments in innovations are pursued. Regarding many patents in renewable energy – particularly in solar – the effects and costs of modern renewable energy, presumably, improve in the future.

7.6 Enhancing Valorisation of Energy Services

Renewable energy would reduce the carbon content in energy resources; however, the decarbonisation of energy resources, measured by CO_2 emissions, is hardly observed. While the carbon content per GDP declined in many countries which was largely due to faster growth of GDP than consumed energy, globally, the carbon per capita and per energy unit hardly changed. They declined in high-income countries but increased in most mid-income and low-income ones that turned from traditional renewable energy to fossil fuels. The carbon content of energy consumption increased which even accelerated during the years of high fuel prices because several mid- and low-income countries turned from oil and gas to cheaper coal. This decarbonisation diverged across countries, making it laborious for international policies to encourage CO_2 emission reduction.

Massive afforestation, agroforestry, nature areas, and suchlike methods absorb CO_2 from the air but the sequestration declines. Hydrogen production is pursued when aiming at decarbonisation. However, the present production of hydrogen is based on methane in natural gas, which hardly reduces CO_2 because uses a lot of energy and generates methane in the life cycle. Capture and storage of CO_2 hardly help because needs more energy and costs. Similar can be expected for the pyrolysis of natural gas. These methods consume more natural gas, thereby, foster complementation of fossil fuels rather than a substitution for renewable energy. Electrolysis of water for hydrogen is small-scale, despite more than a century of experimentation because needs much electricity and high investment per unit hydrogen. Electrolysis based on renewable energy enables storage of the intermittent energy resources and substitution of some coal at a reasonable CO_2 price, but the decarbonisation of energy consumption in the next decades is an illusion as it is small scale and even plans for the next decades are insufficiently large scale. Moreover, R&D and patents in hydrogen declined during the 2010s whilst the cost-reducing technical change was slow. Therefore, perspectives of innovations in hydrogen are bleak, whereas consumption is risky because hydrogen is explosive.

More promising is the valorisation of energy services. Electricity services with renewable energy add value in consumption, in particular when foster distributed energy system. Therefore, the storage of the intermittence of wind and solar energy is needed, as well as development of effective and cheap systems for storage and distribution of heat because these technologies are deficient and too costly. By 2010, the electricity storage was mainly based on hydro storage, whereas twenty times large storage is needed for the growing energy consumption, growing share of intermittent modern renewable energy in it, and more frequent exchanges of electricity between producers and consumers on the grid and off the grid. By far the largest storage is with hydro applications. In addition, manifold larger capacity of batteries for storage are needed. Furthermore, investments in local, domestic, and international grids are necessary to prevent electricity losses and blackouts, as well as investments in cheap, local distribution and storage of residual heat.

The valorisation of energy services evolved from 1990 to 2015 which has generated annual value addition. Modern renewable energy contributed to that. The growing value-added in energy services generates sufficient income for innovations in distributed energy systems. As the valorisation of energy services is observed in nearly all countries, this converging trend across countries fosters changes toward renewable energy in the next decades. The protectionist argument that regulations of fossil fuel causes ‘outshoring’ is not valid because the valorisation of energy services evolved in nearly all OECD and non-OECD countries. Extrapolations of the past trends indicate that higher income, accessible energy along with far-reaching CO_2 emission reduction is possible under the condition that policies support the substitution of fossil fuels for renewable energy. If the annual average growth rates of income, energy consumption and renewable energy during 1990–2015 continue unchanged during 2015–2040, global income grows nearly four times, particularly in mid- and low-income countries, and energy consumption doubles mainly in low-income countries whilst global CO_2 emissions are reduced to 46% of the emissions in the 2015 year because renewable energy increases more hundred times, mainly in China, India, Ethiopia and the EU. Such emission reduction enables the mitigation of climate change. Many issues are going to be faced, but they are solvable through better designs and applications as shown in various pilot projects.

A prosperous, energy-consuming world along with far-reaching CO_2 emission reduction is attainable if people pursue higher value energy services rather than maintain habitual fossil fuels while the authorities resist protectionist pressures of incumbent interests but support sustainable innovations.

7.7 Summary Mechanisms of Change

Main mechanisms of change observed in the past are summarised below (Table 7.1).

Table 7.1 Summary of the mechanism of change in energy production and consumption

Impediments for renewable energy	Enhancements for renewable energy
Complementation and diversification	
Complementation adds resources to each other rather than substitutes them. Divergence in the carbon content of energy consumption across countries.	Shift from agriculture and industry to service decouples income from energy and CO_2. Diversified energy resources enable effective energy services. Diversified energy services enable value addition and more income from energy consumption.
Changing economies	
More efficient energy production adds welfare and energy access, but the rebound effect in consumption counter positive changes. Large interests in fossil fuels oppose changes to novel renewable energy.	Increase in fuel prices invoked new renewable energy resources, though not per se the growth of novelties. Engineering capabilities generate the effect-increasing and cost-reducing technical change over time. Functional and ethical qualities in the energy services generate higher income per consumed energy.
Innovating in renewable energy	
It takes 6–9 decades from novelty to diffusion of basic energy technologies, shorter if driven by large public interests but longer if environmental concerns arise. Chances that assessors select a successful R&D project measured by patents and start-up is close to nil. Chances that an excellent R&D generate surviving startups are close to nil.	R&D in renewable energy and related areas generated patents effectively and efficiently. Societal and far-reaching policy demands increase manifold the chance that an R&D project foster competitive startups.
Growth of renewable energy	
Policy support of fossil fuels effectively impedes renewable energy Inconsistent subsidy policy and negligible CO_2 tax obstruct welfare and progress in renewable energy	Emission cap with emission trading enables the dissemination of renewable energy. Shift of policy support from fossil fuels to value-adding energy services is cost-effective and fair. Guaranteed purchases of renewable energy provide stability on market. Stakeholder participations fosters continuity in investment, thereby cost-reducing changes.
Decarbonisation and Valorisation	
Substitutions of traditional renewable energy for the modern ones instead of fossil fuels. Decarbonisation of energy resources diverged across countries, which impedes international policies.	Valorisation of energy services converged across countries which fosters policies on energy access and emission reduction.

Appendices

Appendix 1: Calculation Methods

The formulas used in the book are shown and explained below. The value added is the sale of products minus the purchase of resources. It shows deliveries of capital, labour and knowledge in an organisation, excluding purchases of materials, energy and hired labour, formally represented as:

$$V_i = (P_{s,i} - P_{r,i}) \cdot Q_i \tag{A.1}$$

Where 'V_i' is value added in monetary terms; 'P_s' is the sales price; 'P_r' is the resource price; 'Q_i' denotes the quantity of resources in terms of mass; for each resource, purchase prices are multiplied by quantities.

Wherein historical changes are analysed, current prices of consumer goods are converted into the constant prices of a year i.e., real prices. Estimates of constant prices 'p_r' are corrections of current prices with the inflation coefficients 'r' in databases. If uncorrected in the databases, the World Bank inflation rates are used for the most part of price estimates, formally represented as:

$$p_r = p_t / \left(\frac{r}{100} + 1\right) \tag{A.2}$$

The prices corrected for inflation are expressed in the money value of a particular year; for instance, US_{2011} means prices in US Dollars as per the year 2011. Note that when the annual rate of inflation is high, these corrections are laborious and somewhat arbitrary because they refer to the prices of traded goods rather than the prices of consumer goods on the domestic market.

Changes are called 'growth'. Growth 'g_t' is assumed to be an exponential function of time 't' per year, formally represented as:

Y. Krozer, *Economics of Renewable Energy*,
https://doi.org/10.1007/978-3-030-90804-1

$$g_t = {}^{n_{t+1}}\!/_{n_t} - 1 \tag{A.3}$$

The growth rate 'g' is often an annual average of several years over a period of time. High growth rates generate large changes within a few years, but a low rate spanning a long period of time also causes large changes; for example, twice the volume is generated in 70 years with 1% growth $(1+0.01)^{70}$; in 36 years with 2% growth; and in 24 years with 3% growth. Herewith, low rates are relevant because long-time spans are assessed.

Trends in the growth rate during a period are called accelerators, 'a', formally represented as:

$$a_t = {}^{g_{t+1}}\!/_{g_t} - 1 \tag{A.4}$$

A trend above 1 implies acceleration of growth; and below 1 a slow-down. Acceleration over a few subsequent decades is an augmenting trend; whereas, the slow-down over a few decades indicates saturation. For instance, augmenting renewable energy is an acceleration of the growth, but a saturating population means slow-down of the growth.

The trends encompass fluctuations, such as booms and slumps of income. Herewith, the standard deviation 's' is used to indicate the scale of fluctuations. This is formally represented as:

$$\mathrm{s} = \sqrt{\left[{}^{1}\!/_{(N-1)}\right] \cdot \sum_{i=1}^{N} (x_1 - \bar{x})^2} \tag{A.5}$$

It is calculated in Excel. A large standard deviation indicates strong fluctuations. Increasing standard deviation spanning a period of time indicates a diverging trend, and a decreasing one shows a converging trend; for example, divergence in CO_2 emissions per capita across countries and convergences in energy consumption. The standard deviation, when compared to the average, is also used as an indication of the validity of conclusions; when the standard deviation is higher than the average, the validity of conclusion is low. For example, it can be a disputable statement with respect to large fluctuations.

Links between observations are correlated; for example, the changes in income when compared to changes in energy consumption. The Pearson correlation is used because it is a reliable and well-known method that is easy to handle in Excel by the interested public. A regression is formally represented as:

$$R^2(x, y) = {}^{\sum (x-\bar{x}) \cdot (y-\bar{y})}\!\Big/_{\sqrt{\sum (x-\bar{x})^2 \cdot \sum (y-\bar{y})^2}} \tag{A.6}$$

For example, 'x', is the annual average growth of income and, 'y', is the annual average growth of energy consumption during a period. Herewith, a stringent criterium is used because regressions equal to and above 0.8 ($R^2 \geq 0.8$) are considered correlated, and between 0.5 and 0.8 moderately correlated but lower are assumed uncorrelated.

A mainstream assumption is that higher prices reduce demands. The price elasticities of demand refer to the effects of price changes on demand changes; for instance, the effect of higher prices of energy resources on the consumed volume of energy. The mid-price elasticities are calculated and formally represented as:

$$k = \frac{p_{i+1} - p_i}{(p_{i+1} + p_i)/2} / \frac{q_{i+1} - q_i}{(q_{i+1} + q_i)/2} \tag{A.7}$$

For $p_i = \sum_i^n p_t/n$, also for q_i; $p_{i+1} - p_i$ is price change, $q_{i+1} - q_i$ volume change

In energy, low price elasticity '*e*' is usually assumed ($e > -1$), meaning that an increase of the unit price causes less than a unit consumption decrease. The income elasticity is estimated in a similar way for the incomes instead of prices. In energy, income elasticity above +1 is usually assumed; it means that a unit of additional income causes more than a unit of additional consumption.

The energy-efficiencies of conversions usually increase over time, called the effect-increasing technical change. It is expressed in percentage annual change, formally represented as:

$$k_t = k_0 a^t \tag{A.8}$$

Herewith, 'k_t' is the effect due to the change; 'k_0' the initial effect; 'a' the coefficient of the effect-increasing technical change usually assumed due to know-how and scale of activities over time 't', usually in years.

The costs of technologies often decrease over time. This cost-reducing technical change is expressed in a percentage average change of unit cost per year, formally represented as:

$c_t = c_0 a^t$

The unit cost, 'c', also called 'marginal cost', is total cost divided by volume in a year. A popular measurement is the unit costs per double output, formally shown as:

$$c_t = c_0 \left(\frac{o_t}{o_0}\right)^b \tag{A.9}$$

Wherein 'c_t' denotes the unit cost in year t; 'c_0' the unit cost in base year; 'o_t' the output in year t; 'o_0' output in the base year; and 'b' the inclination of the curve that reflects the cost-reducing technical change. The cost reduction is measured by half-time, which is the time needed for the reduction of unit costs by half; it is $c_t/c_0 = 0.5$.

This is much used to estimate the increases in the required scale of the production, given the cost reducing technical change.

In engineering, the effect-increasing change is often used to indicate changes in energy efficiency; and in economics, the cost-reducing changes are used to indicate impacts of technological changes on prices. These two are often combined and called 'technological learning'; whilst the effect-increasing technical change always reduces some costs, the cost changes may not influence the effect when cheaper resources are used without further effect.

The conventional indicator of business performance is profitability. This is estimated with the Net Present Value (NPV). Firstly, the Present value is calculated followed by the Net Present Value:

$$PV_t = \sum_{t=1}^{n} \mathrm{C}/(1 + r/b)^{bt} \tag{A.10}$$

For example, if 1 million USD is invested (also called the outgoing cash flow) 'I' and 1 million USD of income is generated per year during two years (also called the incoming cash flow) 'C' at 10% interest 'r', the NPV is:

1st year $1/(100\%+10\%)^1 = 0.91$
2nd year $1/(100\%+10\%)^2 = 0.83$ million Euro
PV $= 1.74$
$NPV = PV - I$
NPV is $1.74 - 1 = +0.74$

Positive NPV indicates potentially profitable investment, negative NPV a loss. When several investments over time are done, they should be discounted much in the same way as the income. If alternative proposals with different timelines are compared, the present value of average annual cash flows are estimated called the 'uniform annual stream' (UAS), 'equivalent annual cost' (EAC), 'present value annuity', or suchlike names. These terms are often used to express annual capital costs of investments. It is calculated as the present value or net present value divided by the annuity factor. The annuity factor A_t is the sum of the discount factors over time, and the uniform annual streams (UAS) is NPV (or PV for nil investment) divided by A_t:

$A_t = \sum_t^n 1/(100\% + r)^t$
$UAS = {}^{NPV}/_{A_t}$

Based on the above: $A_t = (100\% + 10\%)^1 + (100\% + 10\%)^2] = 1.1 + 2.21 = 3.31$
UAS $= 0.74/3.31 = 0.22$ average profit a year.

Appendix 2: Countries Income and Energy

Table A.1 covers: countries' GDP and energy production per capita in the years 1900 and 2000; as well as the annual average growth in ten-year intervals based on the average of the preceding ten years, and correlations between GDP and TWh production in these countries. Highlighted in bold are the net exporters of energy, being larger producers than consumers.

Table A.1 Countries' income and energy indicators; () means negative number

Data Excluding a Few Extremes in the 1900s	GDP USD/cap in 2000	Energy kWh/Cap in 2000	USD/ kWh in 1900 or Earliest	USD/ kWh in 2000	1900–1950 growth	1960–2000 growth	Correl GDP: TWh
1. Argentina	14,918	28,673	46	0.5	−7%	−3%	0.33
2. Australia	36,001	170,035	0.4	0.2	1%	−3%	(0.07)
3. Austria	35,714	19,109	0.1	1.9	3%	3%	0.87
4. Belgium	33,427	14,192	0.2	2.4	0.4%	4%	0.74
5. Bolivia	2984	17,084	43	0.2	−11%	−4%	0.51
6. Brazil	8316	12,905	35	0.6	−6%	−2%	0.20
7. Bulgaria	8671	14,097	11.1	0.6	−6%	1%	0.37
8. Canada	37,446	174,462	0.4	0.2	−0.5%	−1%	0.09
9. Chile	10,903	6422	1.2	1.7	−1%	1%	0.37
10. China	2460	16,288	10	0.2	−6%	−2%	0.69
11. Colombia	6860	25,115	0.8	0.3	−2%	−1%	0.47
12. Cuba	4882	3888	83	1.3	−4%	−7%	0.15
13. Czech & Slovakia	16,153	29,050	0.4	0.6	−1%	2%	0.69
14. Denmark	35,923	63,797	4.3	0.6	17%	−3%	0.86
15. Germany	33,975	18,114	0.3	1.9	−1%	4%	0.86
16. Egypt	5269	13,824	35	0.4	−7%	−3%	0.59
17. Finland	32,972	25,424	3.1	1.3	−3%	−0.1%	0.29
18. France	31,771	24,628	0.6	1.3	−0.2%	2%	0.64
19. Ghana	2337	1054	1.7	2.2			N.A.
20. Greece	23,138	10,111	71	2.3	−4%	−3%	0.42
21. Hungary	14,757	11,279	0.3	1.3	−0.3%	3%	0.41
22. India	2003	3083	5.1	0.6	−2%	−2%	0.74
23. Indonesia	3472	12,312	3.6	0.3	−3%	−1%	0.70
24. Iran	7573	53,767	2.1	0.1	−6%	−0.04%	(0.04)
25. Iraq	3147	55,066	3.6	0.1	−13%	1%	0.41
26. Ireland	39,152	2902	21	13	−7%	6%	0.94
27. Italy	33,185	6185	27	5.4	−6%	3%	0.63
28. Jamaica	5342	561	100	9.5	−9%	−1%	0.11
29. Japan	33,294	9574	0.8	3.5	−1%	4%	0.94
30. Jordan	3896	662	16	5.9		−5%	(1.00)

(continued)

Table A.1 (continued)

Data Excluding a Few Extremes in the 1900s	GDP USD/cap in 2000	Energy kWh/Cap in 2000	USD/ kWh in 1900 or Earliest	USD/ kWh in 2000	1900–1950 growth	1960–2000 growth	Correl GDP: TWh
31. Korea	22,930	13,491	25	1.7	−7%	2%	0.78
32. Lebanon	11,232	536	36	21	−4%	−0.2%	0.09
33. Mexico	11,338	29,254	2.9	0.4	−3%	−0.2%	(0.55)
34. Morocco	4580	165	34	28	−9%	3%	0.18
35. Myanmar	999	2890	1.1	0.3	6%	−5%	0.25
36. Netherlands	39,923	49,412	5	0.8	−4%	1%	(0.16)
37. New Zealand	25,868	49,389	1.0	0.5	0%	−1%	0.63
38. Norway	54,594	672,999	7.1	0.1	−9%	−3%	0.72
39. Peru	4777	5326	1.2	0.9	−2%	1%	(0.15)
40. Philippines	4187	1301	91	3.2	−3%	−2%	0.89
41. Poland	13,207	20,743	0.2	0.6	−1%	3%	0.71
42. Portugal	21,497	4054	90	5.3	−7%	1%	0.71
43. Romania	7430	15,116	0.9	0.5	−5%	4%	0.02
44. Saudi Arabia	20,129	330,486	3.0	0.1	−20%	1%	0.44
45. South Africa	8478	38,859	0.4	0.2	−0.1%	−1%	0.34
46. CIS	11,532	50,627	1.2	0.2	−2%	−1%	(0.20)
47. Spain	26,424	10,284	3.2	2.6	−2%	2%	0.90
48. Sri Lanka	4391	499	29	8.8		−2%	0.85
49. Sweden	36,374	67,085	7.9	0.5	−4%	−1%	0.46
50. Switzerland	42,752	24,985	1.3	1.7	−1%	1%	0.40
51. Syria	4946	23,334	40	0.2		−11%	0.34
52. Taiwan	31,937	13,080	3.3	2.4	−5%	5%	0.89
53. Thailand	6921	7836	131	0.9		−9%	(0.23)
54. Tunisia	8440	7925	32	1.1		−6%	N.A.
55. Turkey	11,830	4493	5.6	2.6	−3%	2%	0.33
56. United Kingdom	23,696	45,814	0.1	0.5	2%	1%	0.35
57. United States of America	45,887	71,283	0.2	0.6	1%	2%	0.29
58. Uruguay	6269	8929	2.9	0.7	3%	−3%	0.66
59. Venezuela	7927	100,449	5.8	0.1	−12%	3%	0.42
60. Vietnam	2792	29,030	6.8	0.1	−2%	−6%	0.49
61. Yugoslavia	15,524	13,359	3.1	1.2	−3%	1%	0.44
World Total	10,814	27,751	0.4	0.4	−1%	0.5%	0.24

Appendix 3: Success Rates of Innovations in EU (Table A.2)

Table A.2 Rates of successful innovation processes across the EU countries in all sectors (ALL) and in the broadened renewable energy sector (RES+)

Assumed USD_{2010} 0.5 million /R&D project	ALL: all firms				RES+			
		Development performance	Diffusion performance			Development performance	Diffusion performance	
	Number of R&D projects	Patent/R&D projects	Survivor / Birth firms	Success rate of R&D	Number of R&D projects	Patent/R&D projects	Survivor / Birth firms	Success rate of R&D
EU	699,209	8%	10%	1%	9931	18%	74%	13%
Belgium	22,787	6%	44%	2%	356	9%	93%	8%
Bulgaria	715	3%	20%	1%	8	21%	59%	12%
Czech	6940	4%	25%	1%	85	6%	122%	8%
Denmark	19,693	6%	11%	1%	322	53%	−3%	−2%
Germany	203,003	12%	−8%	−1%	1834	39%	68%	27%
Estonia	746	5%	18%	1%	53	3%	−13%	−0%
Ireland	7391	5%	50%	2%	92	11%	116%	13%
Greece	3982	2%	−18%	−0%	10	35%	33%	11%
Spain	35,281	4%	−11%	−0%	295	28%	249%	69%
France	118,989	8%	36%	3%	2626	10%	81%	8%
Croatia	950	N.A.	−8%	N.A.	22	N.A.	40%	N.A.
Italy	53,879	8%	−9%	−1%	982	10%	107%	11%
Cyprus	228	4%	−26%	−1%	N.A.	N.A.	74%	N.A.
Latvia	342	4%	27%	1%	N.A.	N.A.	57%	N.A.
Lithuania	784	3%	21%	1%	15	N.A.	50%	N.A.
Luxembourg	1634	5%	30%	1%	N.A.	N.A.	73%	N.A.
Hungary	3361	6%	−4%	−0%	142	2%	31%	1%
Malta	136	8%	15%	1%	N.A.	N.A.	73%	N.A.
Netherlands	31,850	9%	27%	3%	438	16%	35%	6%
Austria	23,731	8%	2%	0%	304	15%	42%	6%
Poland	8465	5%	11%	1%	174	7%	29%	2%

(continued)

Table A.2 (continued)

Assumed USD_{2010} 0.5 million /R&D project	ALL: all firms				RES+			
		Development performance	Diffusion performance			Development performance	Diffusion performance	
	Number of R&D projects	Patent/R&D projects	Survivor / Birth firms	Success rate of R&D	Number of R&D projects	Patent/R&D projects	Survivor / Birth firms	Success rate of R&D
Portugal	6371	2%	−8%	−0%	12	46%	86%	39%
Romania	1759	3%	33%	1%	66	3%	37%	1%
Slovenia	2088	5%	23%	1%	111	1.5%	76%	1%
Slovakia	1430	3%	27%	1%	41	5%	83%	5%
Finland	17,145	8%	8%	1%	551	7%	64%	4%
Sweden	35,035	7%	29%	2%	363	13%	180%	23%
United Kingdom	90,490	47%	15%	7%	1022	100%	61%	61%

Table A.3 Carbon intensity, carbon performance and carbon efficiency in the EU

	Carbon intensity in tonne CO_2 per capita		Carbon performance in €$_{2010}$ per kg		Carbon efficiency in kg CO_2 / kWh	
	2018	Growth 2008–2018	2018	Growth 2008–2018	2018	Growth 2008–2018
EU28	5.5	−1.9%			0.15	−1.1%
Belgium	6.8	−2.3%	10.7	−1.4%	0.12	−1.0%
Bulgaria	5.8	−2.4%			0.18	−1.8%
Czechia	8.1	−1.3%	5.3	−3.5%	0.17	−1.4%
Denmark	11.8	−2.0%	7.1	−2.9%	0.33	−1.0%
Germany	7.5	−2.7%	9.0	−2.2%	0.17	−0.6%
Estonia	13.0	−1.5%	2.5	−1.6%	0.24	−1.0%
Ireland	8.9	0.8%			0.25	4.2%
Greece	5.5	2.7%	5.3	−0.5%	0.21	−1.3%
Spain	4.6	−3.6%			0.14	−1.5%
France	3.4	−2.3%			0.08	−1.3%
Croatia	3.1	−2.3%			0.13	−2.0%
Italy	4.2	−2.8%			0.14	−2.2%
Cyprus	6.9	−3.8%			0.20	−1.3%
Latvia	4.0	−3.1%	5.8	−0.5%	0.14	−1.1%
Lithuania	5.0	0.5%			0.15	3.2%
Luxembourg	12.6	2.2%	21.5	−4.9%	0.15	0.8%
Hungary	3.8	−1.7%	7.0	−2.8%	0.12	−1.5%
Malta	7.4	−1.3%			0.36	2.2%
Netherlands	8.5	−0.7%			0.16	−0.1%
Austria	5.5	−1.1%			0.12	−1.0%
Poland	7.8	−1.7%			0.24	−0.8%
Portugal	4.2	0.1%			0.15	−0.9%
Romania	3.2	−1.2%			0.16	−2.4%
Slovenia	6.2	−3.2%			0.16	0.0%
Slovakia	5.5	−1.7%			0.15	−0.7%
Finland	8.0	−1.5%	8.8	−3.2%	0.11	−1.7%
Sweden	3.8	−2.2%	21.5	−.26%	0.07	−1.9%
Un. Kingdom	4.6	−2.5%			0.01	0.7%

Appendix 4: Checklist Possible Benefits

Possible benefits of energy services with renewable resources in the literature

	Producers' attributes	Consumers' attributes
Individual	• Resource diversification Local resources can be identified and utilized (NREL, 1997; Resch et al., 2008; EPA, 2011) Financial risks in resource purchase can be spread (Awerbuch, 2004; Adamec et al., 2011) • Management Congestion on grid can be relieved with flexible production (EPA, 2011; Richter, 2012; Eyer & Corey, 2010; Castillo & Gayme, 2014; Burger & Luke, 2016) Attention for stability of voltage is invoked (Eyer & Corey 2010; Adamec, 2011; EPA, 2011; Richter, 2012; Castillo & Gayme, 2014) Back up and reserves can expand to reduce power losses (Schiermeier et al., 2008; EPA, 2011; Richter, 2012; Castillo & Gayme, 2014; Siao, 2014; Bronski et al., 2015) Lower peak costs when production is more flexible (Eyer & Corey, 2010; EPA, 2011; Richter, 2012; Busch & McCormick, 2014; Neubauer & Simpson, 2015) • Marketing Price volatility is mitigated when renewables are add-on (Awerbuch, 2004; Adamec et al., 2011; EPA, 2011) Certification of the sustainability performance can reduce regulation (Schiermeier et al., 2008; Richter, 2012) Branding of renewable energy enables new tariffs (Adamec et al., 2011) Flat rate fee due to renewables add-on is easier to manage (Schleicher-Tappeser, 2012) Diverse renewable energy justifies the dynamic pricing (Schleicher-Tappeser, 2012; Eid et al., 2016) • Change of current Distributed systems foster cheaper supply of direct current (Pepermans et al., 2003) Cheaper low voltage infrastructure is possible (Levin & Thomas, 2016)	• Prices Monopoly price of suppliers are mitigated (Pepermans et al., 2003; Schleicher-Tappeser, 2012; Eid et al., 2016) Peaks in supplies can be stored in appliances (Adamec et al., 2011; Neubauer & Simpson, 2015) Pricing can be based on local nodes rather than generic (Schleicher-Tappeser, 2012) Charges on the decreasing demands can be avoided (Richter, 2012) • Volume Electricity from waste heat can be generated (Pepermans et al., 2003; Rismanchi, 2017) Incentive to reduce peaks are easier to introduce (Tuballa & Abundo, 2016) Appliance can be integrated into a system (Adamec et al., 2011; Tuballa & Abundo, 2016) Consumption of energy can be better monitored (Adamec et al., 2011) Individual and collective autonomy increase (NREL, 1997; Pepermans et al., 2003; Richter, 2012; Siao, 2014; Eid et al., 2016)

(continued)

	Producers' attributes	Consumers' attributes
Collective	• Mitigation of impacts External effects on environment are reduced NREL, 1997; Schiermeier et al., 2008; EPA, 2011; Eller & Gauntlett, 2017) Import dependency is reduced (Resch et al., 2008; Eller & Gauntlett, 2017) • Local stakeholders Local constructors' capabilities are developed (NREL, 1997) Local business in electricity storage is generated (Schiermeier et al., 2008; EPA, 2011; Eller & Gauntlett, 2017) Polluted land and wasteland can be used (Busch & McCormick, 2014) Wasted energy can be recovered (EPA, 2011; Rismanchi, 2017) Use of renewable resources can be optimised (Rismanchi, 2017) Large-scale, costly investment can be deferred (Adamec et al., 2011; EPA, 2011; Bronski et al., 2015; Burger & Luke, 2016) Energy consumption in the remote areas is enhances (Eller & Gauntlett, 2017)	• Economy Larger local income in generated (Richter, 2012; Busch & McCormick, 2014) More local jobs are created (NREL, 1997; Busch & McCormick, 2014) Higher and more stable tax income is generated (NREL, 1997; Busch & McCormick, 2014) Better regional image attracts business (Richter, 2012; Busch & McCormick, 2014) Diversification enables price stability (Busch & McCormick, 2014) Diversification reduces investment peaks (Busch & McCormick, 2014 • Social interest Cumulative benefits are generated (Adamec et al., 2011) Isolated consumers can be better served (Akinyele & Rayudu, 2014) Social capabilities increase (Busch & McCormick, 2014) Services can be tuned to demands in communities (Tuballa & Abundo, 2016)

Appendix 5: Carbon Intensity, Performance, Efficiency

Table 1 Carbon intensity, carbon performance and carbon efficiency in the EU						
	Carbon intensity in tonne CO_2 per capita		Carbon performance in €$_{2010}$ per kg		Carbon efficiency in kg CO_2 / kWh	
	2018	Growth 2008-2018	2018	Growth 2008-2018	2018	Growth 2008-2018
EU28	5.5	-1.9%			0.15	-1.1%
Belgium	6.8	-2.3%	10.7	-1.4%	0.12	-1.0%
Bulgaria	5.8	-2.4%			0.18	-1.8%
Czechia	8.1	-1.3%	5.3	-3.5%	0.17	-1.4%
Denmark	11.8	-2.0%	7.1	-2.9%	0.33	-1.0%
Germany	7.5	-2.7%	9.0	-2.2%	0.17	-0.6%
Estonia	13.0	-1.5%	2.5	-1.6%	0.24	-1.0%
Ireland	8.9	0.8%			0.25	4.2%
Greece	5.5	2.7%	5.3	-0.5%	0.21	-1.3%
Spain	4.6	-3.6%			0.14	-1.5%
France	3.4	-2.3%			0.08	-1.3%
Croatia	3.1	-2.3%			0.13	-2.0%
Italy	4.2	-2.8%			0.14	-2.2%
Cyprus	6.9	-3.8%			0.20	-1.3%
Latvia	4.0	-3.1%	5.8	-0.5%	0.14	-1.1%
Lithuania	5.0	0.5%			0.15	3.2%
Luxembourg	12.6	2.2%	21.5	-4.9%	0.15	0.8%
Hungary	3.8	-1.7%	7.0	-2.8%	0.12	-1.5%
Malta	7.4	-1.3%			0.36	2.2%
Netherlands	8.5	-0.7%			0.16	-0.1%
Austria	5.5	-1.1%			0.12	-1.0%
Poland	7.8	-1.7%			0.24	-0.8%
Portugal	4.2	0.1%			0.15	-0.9%
Romania	3.2	-1.2%			0.16	-2.4%
Slovenia	6.2	-3.2%			0.16	0.0%
Slovakia	5.5	-1.7%			0.15	-0.7%
Finland	8.0	-1.5%	8.8	-3.2%	0.11	-1.7%
Sweden	3.8	-2.2%	21.5	-.26%	0.07	-1.9%
Un. Kingdom	4.6	-2.5%			0.01	0.7%

Table A.3 shows the carbon intensity, carbon performance, and carbon efficiency in the EU28 member countries. Totals in 2018 and annual average growth during 2008–2018 are shown.

Appendix 6: Inputs and Outputs of Hydrogen Production

Data for the Hydrogen Production

Impacts of hydrogen for coal substitution on CO_2 emission reduction and costs in USD_{2019} are estimated. Tables show energy and emission coefficients, input-output data, substitution, emission reduction and cost of four technologies:

1. Steam Methane Reforming with Water Gas Shift (SMR + WGS), about 97% of global production
2. Steam Methane Reforming with Water Gas Shift and Carbon Capture and Storage (SMR + WGS + CCS)
3. Methane Pyrolysis (MP) in the pilot stage
4. Electrolysis of water (EL), about 3% of global production (up to kWh 50–55 per kg H_2) (Tables A.4, A.5, A.6 and A.7).

Table A.4 Assumed coefficients for the estimates

per kg H_2	kWh/kg	CO_2 kg/kWh	CO_2 kg/kg
Hydrogen	33	0	0
Natural gas	14	0.2	2.6
Coal (bisemous)	7	0.3	2.3

Table A.5 Shows input–output data based on the Wikipedia

kWh, kg, TWh, million tonnes	SMR +WGS	SMR+WHG – CCS	Methane Pyrolysis	Electrolysis
Inputs				
Heat kWh	5.7	5.7	5.2	0
Electra kWh	0	0	0	39.4
Methane kg	2.2	2.9	4.4	0
Water kg	4.9	3.3	0	9.1
Outputs				
H_2 kg	1	1	1	1
CO_2 kg	6.0	1.5	0	0
CO_2 storage kg	0	3.6	0	0
O_2 kg	0	0	0	8.0
Carbon kg	0	0	3.3	0
Energy kWh				
In	36	45	65	39
Out	33	33	33	33
Out-In	−2	−12	−32	−6

Table A.6 CO_2 emission reduction due to those substitutions in kg/kg H_2

Coal reduction	11.7	11.7	11.7	11.7
Net CO_2 reduction	5.7	10.2	11.7	11.7
CO_2 equivalent oil	8.5	8.5	8.5	8.5
Net CO_2 reduction	2.5	7.0	8.5	8.5
CO_2 equivalent gas	7	7	7	7
Net CO_2 reduction	1.0	5.5	7.0	7.0

Table A.7 Costs and emission reduction (all costs in USD_{2019})

Electra	0.2	0.2	0.2	1.5
Power gas	0.4	0.4	0.7	0
CCS	0.0	0.5	0	0
Total	0.6	1.1	0.9	1.5
Cost-effectiveness estimates				
CO_2 reduction for coal	49%	87%	100%	100%
Cost-effect USD_{2019}/tonne	103	106	80	128
CO_2 reduction for gas	14%	78%	100%	100%
Cost-effect USD_{2019}/tonne	606	198	135	215

Appendix 7: Indicators of the Valorisation

Table A.8 shows the purchasing power per capita in 2015, and its growth based on the World Bank data. The countries with more accessible energy due to income growth and declining disparities are shown bold. Income disparities within countries are shown with Gini coefficients in the years of the earliest estimate in the database and in 2015. Gini coefficients indicate division of income in ten equal income classes indexed from 0 to 1; 0 being no income difference and 1 the maximum difference in two categories: extremely poor and rich. Although the Gini coefficients are criticised because not cover all countries and wealth groups, do not consider the division between wages and capital, do not specify the highest few percent of income which covers a large part of all income; yet, they are widely used.

Table A.9 shows the energy indicators in 2015, and the average growth in the period from 1990 to 2015; the growth rates during 1990–2005 and 2005–2015 are also estimated, and discussed below; they are not presented in this table. The indicators are: energy performance, energy and electricity consumption, shares of renewable energy and modern renewable energy in energy consumption.

Table A.8 Incomes per capita: annual average growth and disparities

	GDP-PPP		Gini coefficients	
Bold: growing accessibility	USD per capita	Growth 1990–2015	Earliest found	2015
World	15,694	**4.4%**	N.A.	N.A.
USA	56,469	**3.5%**	38	41
Japan	40,607	**3.0%**	32	–
EU	38,447	**3.9%**	N.A.	N.A.
Russia	24,692	**5.5%**	48	**38**
Mexico	16,983	**4.6%**	50	**46**
Brazil	15,617	**3.8%**	61	**51**
China	14,450	**11.5%**	43	**42**
Indonesia	11,040	**5.5%**	40	–
Philippines	7320	**4.2%**	N.A.	N.A.
India	6127	**7.0%**	35	–
Nigeria	6038	**4.9%**	45	**43**
Pakistan	5000	**3.8%**	33	34
Bangladesh	3336	**5.7%**	28	32
Ethiopia	1633	**5.6%**	45	**39**

Table A.9 Indicators for assessing valorisation of energy consumption: energy performance, energy and electricity consumption, share of all renewable energy and modern renewable energy in 2015 and annual average growth

	Energy performance		Energy consumption		Electricity consumption		Share renewables in consumption		Share modern renewables in consumption	
	USD/ kWh	Growth	kWh/ cap	Growth	kWh/ capita	Growth	Share	Growth of share	Share	Growth of share
Growth is annual average during 1990–2015	2015	1990–2015	2015	1990–2015	2015	1990–2015	2015	1990– 2015	2015	1990– 2015
World	0.7	1.6%	21,642	0.5%	3052	1.1%	14%	0.2%	1.5%	5.2%
USA	0.2	0.2%	79,107	−0.4%	12,833	0.9%	7%	1.1%	1.4%	2.8%
Japan	0.9	1.0%	39,359	−0.2%	7864	0.8%	6%	2.5%	1.5%	3.2%
EU	0.9	1.7%	36,205	−0.4%	5968	1.0%	14%	4.6%	2.9%	11.7%
Russia	0.4	1.4%	57,278	−0.7%	6588	0.7%	3%	0.2%	0.0%	12.1%
Mexico	0.9	1.1%	17,974	0.3%	2231	2.4%	8%	−1.3%	2.2%	−1.5%
Brazil	0.9	−0.3%	16,606	1.7%	2506	0.5%	41%	−0.6%	0.9%	18.3%
China	0.5	4.7%	25,218	4.3%	4047	4.2%	9%	−3.8%	1.6%	31.6%
Indonesia	1.0	1.5%	10,176	1.9%	823	4.8%	33%	−1.3%	7.7%	5.5%
Philippines	1.1	1.9%	6061	0.5%	749	2.5%	59%	−1.9%	18.4%	0.4%
India	0.7	2.3%	7550	2.5%	859	2.2%	25%	−2.4%	0.6%	24.0%
Nigeria	0.6	2.1%	8898	0.4%	144	1.9%	80%	0.1%	0.0%	0.0%
Pakistan	0.8	0.86%	5781	0.89%	488	1.4%	38%	−0.9%	0.1%	0.0%
Bangladesh	1.1	0.9%	2738	2.7%	326	5.1%	25%	−3.1%	0.0%	0.0%
Ethiopia	0.3	3.3%	5850	0.2%	86	5.3%	94%	−0.1%	0.0%	−3.1%

Literature

Adamec, M, Pavlatka, P., & Stary, O. (2011). *Costs and benefits of smart grids and accumulation in Czech distribution system*, ICSGCE 2011: 27–30 September 2011, Chengdu, China

Ahaerne, J. F. (2011). Prospects for the nuclear energy. *Energy Policy, 33*, 572–580.

Ahrens, W. A., & Sharma, V. R. (1997). Trends in natural resource commodity prices: Deterministic or stochastic? *Journal of Environmental Economics and Management, 33*, 59–74.

Akerlof, G. A. (1970). The market for "lemons": Quality uncertainty and market mechanism. *The Quarterly Journal of Economics, 84*(3), 488–500.

Akinyele, D. O., & Rayudu, R. K. (2014). Review of energy storage technologies for power networks. *Sustainable Energy Technologies and Assessment, 8*, 74–91.

Al Musa, S. A., & Khidzir, N. Z. (2018). Critical review of intellectual property of its issue and challenge. *Asian of Public Administration and Law, 1*(1), 9–17.

Aljazeera. (2020). https://interactive.aljazeera.com/aje/2020/saving-the-nile/index.html?mc_cid=39b6440fe8&mc_eid=5ae3a1dd3f. Accessed 4 Apr 2020.

Allen, R. (2012). Backward into the future, The shift to coal and implications for the next energy transition. *Energy Policy, 50*, 17–23.

Amran, U. I., Ahmad, A., & Othman, M. R. (2017). Life cycle assessment of simulated hydrogen production by methane steam reforming. *Australian Journal of Basic and Applied Sciences, 11*(3), 43–50.

APGA. (2020). American Public Gas Association. *A brief history of natural gas*. https://www.apga.org/apgamainsite/aboutus/facts/history-of-natural-gas. Accessed 17 Feb 2020.

Apostolides, A., Broadberry, S., Campbell, B., Overton, M., & Leeuwen, B. (2008). *English agricultural productivity 1250–1850: Some preliminary estimates.* Reconstructing the national income of Britain and Holland, c 1270/1500 to 1850.

Arentsen, M., & Bellekom, S. (2014). Power to the people: Local energy initiatives as seeds of innovations? *Energy, Sustainability and Society, 4*(2), 2–12.

Arentsen, M. J., Dinica, V., & Marquart, E. (2001). Innovating innovation policy. In *Rethinking green innovation policy in evolutionary perspective* (Économies et Sociétés, Série Dynamique technologique et organisation, nr 4) (pp. 563–583).

Arora, V. (2014). *Estimates of the price elasticity of the natural gas supply in the United States*. https://mpra.ub.uni-muenchen.de/54232/. Accessed 12 June 2017.

Asplund, R. W. (2008). *Profiting from clean energy*. Wiley.

Awerbuch, S. (2004). *Portfolio-based electricity generation planning: Implications for renewables and energy security* (SPRU). University of Sussex.

Y. Krozer, *Economics of Renewable Energy*,
https://doi.org/10.1007/978-3-030-90804-1

Ayres, R., & Voudouris, V. (2014). The economic growth enigma: Capital, labour and useful energy? *Energy Policy, 64*, 16–28.

Bar-On, Y., Philips, R., & Milo, R. (2018). The biomass distribution on earth, 2018. *PNAS, 115*, 6506–6511.

Battery University. (2021). *Battery statistics.* https://batteryuniversity.com/learn/archive/battery_statistics. Accessed 21 May 2021.

Bayer, P., & Aklin, M. (2020). The European Union Emissions Trading System reduced CO_2 emissions despite low prices. *PNAS, 117*(16), 8804–8812.

Beaudreau, B. C., & Lightfoot, H. D. (2015). The physical limits to economic growth by R&D funded innovation. *Energy, 84*, 45–52.

Beer, C., Reichstein, M., Tomelleri, R., Ciais, P., Jung, M., Carvalhais, N., Rödenbeck, C., Arain, M. A., Baldocchi, D., Bonan, G. B., Bondeau, A., Cescatti, A., Lasslop, G., Lindroth, A., Lomas, M., Luyssaert, S., Margolis, H., Oleson, K. W., Roupsard, O., Papale, D. (2010). Terrestrial gross carbon dioxide uptake: Global distribution and covariation with climate. *Science, 329*(5993), 834–838.

Bergkek, A., & Jacobsson, S. (2003). The emergence of a growth industry; a comparative analysis of the German, Dutch and Swedish wind turbine industries. In J. S. Metcalfe & U. Canter (Eds.), *Change, transformation and development* (pp. 197–227). Physics-Verlag.

Bertoldi, P., & Boza-Kiss, B. (2017). Analysis of barriers and drivers for the development of the ESCO markets in Europe. *Energy Policy, 107*, 345–355.

Bertoldi, P., Boza-Kiss, B., Panev, S., & Labanca, N. (2014). *ESCO market report 2013, JRC science and policy reports*. European Commission, Ispra.

Bhattacharya, M., Paramati, R., Ozturk, I., & Bhattacharya, S. (2016). The effect of renewable energy consumption on economic growth: Evidence from top 38 countries. *Applied Energy, 162*, 733–741.

Block, F., & Keller, M. R. (2008). *Where do innovations come from? Transformations in the U.S. National Innovation System, 1970–2006*, The Information Technology & Innovation Foundation.

Böhme, D., & Dürrschmidt, W. (2007). *Renewable energy sources in figures – National and international development: Status 2007*. Federal Ministry for the Environment, Nature Conservation and Nuclear Safety, Berlin, mimeo.

Bonaventura, M., Ciotti, V., Panzarasa, P., Liverani, S., Lucasa, L., & Latora, V. (2015). Predicting success in worldwide start-up network. *Scientific Reports (2020). 10,* 345. https://doi.org/10.1038/s41598-019-57209-w. Accessed 9 July 2021.

Boon, F. (2012). *Local is beautiful.* Master thesis, University of Utrecht – Geoscience, mimeo.

Boulding, K. E. (1966). *The economics of the coming spaceship earth.* John Hopkins University Press. https://www.semanticscholar.org/paper/The-Economics-of-the-Coming-Spaceship-Earth-Boulding/88cb333c9c7ebb5d852c037372eccc39c12ca691. Accessed 6 May 2020.

Brahic C. (2007). Climate myths: Human CO_2 emissions are too tiny to matter. *New Scientist*, 16-5-2007; https://www.newscientist.com/article/dn11638-climate-myths-human-co2-emissions-are-too-tiny-to-matter/. Accessed 31 July 2020.

Brito, M., & Sousa, T. (2015). World primary energy production & consumption 1900–2010: What can be learned from past trends. In *12th international conference on energy for a clean environment*, July 5–9, 2015, Lisboa, Portugal.

Bronski, P., Creyts, J., Crowdis, M., Doig, S., Glassmire, J., Guccione, L., Lilienthal, P., Mandel, J., Rader, B., Seif, D., Tocco, H., & Touati, H. (2015). *The Economics of Load Defection*. Rocky Mountains Institute, Boulder

Bunger, J. W. (2011). Oil share and tar sand. In D. S. Ginley & D. Cohen (Eds.), *Fundamentals for energy materials and environmental sustainability*. Materials Research Society.

Burger, S. P., & Luke, M. (2016). *Business models for distributed energy resources: A review and empirical analysis*. MIT.

Burrows, L. (2021). *Deaths from fossil fuel emissions higher than previously thought.* Harvard John Paulson School of Engineering and Applied Science. https://www.seas.harvard.edu/news/2021/02/deaths-fossil-fuel-emissions-higher-previously-thought. Accessed 13 Mar 2021.
Busch, H., & McCormick, K. (2014). Local power: exploring the motivation of mayors and key success factors for local municipalities to go for 100% renewable energy. *Energy Sustainability and Society, 4*, 1–15.
Byrne, J. T., Wang, J., & Taminiau, L. M. (2014). The promise of a green energy economy. In J. Byrne & T. Wang (Eds.), *Green Energy Economies*. Routledge.
Carlson, W. B. (2019). *Edison tesla current war ushered electric age.* National Geographic, 27-9-2019. https://www.nationalgeographic.com/history/magazine/2016/07-08/edison-tesla-current-war-ushered-electric-age/. Accessed 9 July 2021.
Castillo, A., & Gayme, D. F. (2014). Grid-scale energy storage applications in renewable energy, integration: A survey. *Energy Conversion and Management, 87*, 885–894.
CEER. (2018). Council of European Energy Regulators. In *Status review of renewable energy and energy efficiency support schemes in EU*, Brussels, data 2009–2017.
Celsius City. (2020). *Thermal energy storage.* https://celsiuscity.eu/thermal-energy-storage/. Accessed 24 May 2021.
Cheng, V. K. M., & Hammond, G. P. (2017). Life-cycle energy densities and land-take requirements of various power generators: A UK perspective. *Journal of the Energy Institute, 90*(2017), 201–213.
Christensen, C. M. (2000). *The innovator's dilemma*. Harper Business.
Ciarreta, A., Espinoza, M. P., & Pizarro-Irizar, C. (2014). Is green energy expensive? Empirical evidence from the Spanish electricity markets. *Energy Policy, 69*, 205–215.
Cipolla, C. M. (1993). *Before the industrial revolution: European society and economy 1000 – 1700*. W.W. Norton & Co.
Clark, G., & Jacks, D. (2007). Coal and the industrial revolution, 1700–1869. *European Review of Economic History, 11*(1), 39–72.
Clements, B., Coady, D., Fabrizio, S., Gupta, S., Alleyne, T., & Sdralevich, C. (2013). *Energy subsidies reform: Lessons and implications*. International Monetary Fund.
Climate Cleanup. (2020). This initiative of entrepreneurs in the Netherlands aims at applications of CO_2. https://climatecleanup.org/partners/. Accessed 19 Nov 2020.
Cloete, S. (2020). Green or blue hydrogen: Cost analysis uncovers which is best for the hydrogen economy. *Energy Post*, EU 9-11-2020. https://energypost.eu/green-or-blue-hydrogen-cost-analysis-uncovers-which-is-best-for-the-hydrogen-economy/. Accessed 5 May 2021.
Coady, D., Parry, I., Sears, L., & Shang, B. (2015). *How large are global energy subsidies?* (Report, WP/15/105). International Monetary Fund.
Colinsk, J. (1996). Why bounded rationality. *Journal of Economic Literature, XXXIV*, 669–700.
Coolman, F. (2021). *Venturing the green transformation?* Social Europe. https://socialeurope.eu/venturing-the-green-transformation. Accessed 16 Aug 2021.
Craig, J. F., Gerali, F., & MacAuley, R. S. (2020). *The history of the European oil and gas industry (1600s–2000s)* (Special Publication, 465, pp. 1–24). Geological Society, London. https://sp.lyellcollection.org/content/465/1/1.1. Accessed 9 Mar 2020.
Creyts, J., Derkach, A., Nyquist, S., Ostrowski, K., & Stephenson, J. (2007). *Reducing U.-S. Greenhouse emissions, how much at what cost*. McKinsey& Company.
Daly, H. E., & Cobb, J. B. (1994). *For the common good* (2nd ed.). Beacon Press.
Demographia. (2021). *World urban areas.* http://demographia.com/db-worldua.pdf. Accessed 22 May 2021.
Deng, Y. Y., Blok, K., & van der Leun, K. (2012). Transition to a fully sustainable global energy system. *Energy Strategy Reviews, 1*, 109–121.
Dismukes, J. P., Miller, L. K., & Bers, J. A. (2009). The industrial life cycle of wind energy electric power generation ARI methodology modelling of life cycle dynamics. *Technology Forecasting & Social Change, 76*, 178–191.
Dixon, L., Porche, I., & Kulick, J. (2003). *Driving emissions to zero*. Rand Corporation.

DOE. (2021). *Global energy storage database.* https://www.sandia.gov/ess-ssl/global-energy-storage-database-home/. Accessed 17 May 2021.

Doelman, J. C., Stehfest, E., van Vuuren, D. P., Tabeu, A., Hof, A. F., Braakhekke, M. C., Gernaat, D. E. H. J., van den Berg, M., van Zeist, W.-J., Daioglou, V., van Meijl, H., & Lucas, P. L. (2020). Afforestation for climate change mitigation: Potential, risks and trade-offs. *Global Climate Change Biology, 26*, 1576–1591.

Dosi, G. (1997). Opportunity incentives and the collective patterns of technological change. *The Economic Journal, 107*, 1530–1547.

Doyne, F. J., & Lafond, F. (2016). How predictable is technological progress? *Research Policy, 45*, 647–665.

Dragoman, M. C. (2014). *Factors influencing local renewable energy initiatives in different contexts.* Master thesis, University Twente, mimeo.

Duffield, J. A., & Collins, K. (2006). Evolution of renewable energy policy, *Choices, 21* (1). https://www.choicesmagazine.org/2006-1/biofuels/2006-1-02.htm. Accessed 31 Feb 2021.

Dyke, J., Watson, R., & Knorr, W. (2021). *Climate scientists: Concept of net zero is a dangerous trap.* Social Europe, 27-4-2021. https://www.socialeurope.eu/climate-scientists-concept-of-net-zero-is-a-dangerous-trap. Accessed 9 July 2021.

EC. (2019). *Energy subsidies and government revenues from energy products.* SWD (2019) 1 final, Brussels.

EC. (2021). https://ec.europa.eu/info/horizon-europe/missions-horizon-europe_en. Accessed 10 Mar 2021.

Edler, J., & Georghiu, L. (2007). Public procurement and innovation – Resurrecting the demand side. *Research Policy, 36*, 949–963.

EEA. (2011). *Efficiency of conventional thermal electricity and heat production.* European Environmental Agency, EN 19.

EIA. (2020). *Cost and performance characteristics of new generating technologies, annual energy outlook.* US Energy Information Administration.

EIA prices. (2017). https://www.eia.gov/electricity/data.php#avgcost is used until 2012, then extrapolated and corrected based on the US Energy Information Administration, Short-Term Energy Outlook Real and Nominal Prices 2017. https://www.eia.gov/outlooks/steo/realprices/. Prices for coal are similar but for oil and gas somewhat different. Accessed 16 May 2021.

EIA total energy. (2021). https://www.eia.gov/totalenergy/data/annual/index.php. Accessed 16 May 2021.

Eid, C., Bollinger, L. A., Koirala, B., Scholten, D., Facchinetti, E., Lillestam, J., & Hakvoort, R. (2016). Market integration of local energy systems: Is local energy management compatible with the European regulation retail competition. *Energy, 114*, 913–922.

Eller, A., & Gauntlett, D. (2017). *Energy storage trends, markets and opportunities.* Energy Sector Management Assistance Program and International Finance Corporation, World Bank Group.

Emission trading. (2021). https://icapcarbonaction.com/en/ets-prices. Accessed 31 Mar 2021.

Encyclopedia.com. (2020). *Tax law and alcohol.* https://www.encyclopedia.com/education/encyclopedias-almanacs-transcripts-and-maps/tax-laws-and-alcohol#:~:text=HISTORY,throughout%20much%20of%20U.S.%20history.&text=The%20states%20also%20impose%20special,state%20and%20local%20sales%20taxes. Accessed 22 June 2020.

Engineering Toolbox. (2020). https://www.engineeringtoolbox.com/fossil-fuels-energy-content-d_1298.html. Accessed 28 Feb 2020.

Enos, J. L. (1962). *The rate and direction of inventive activity: Economic and social factors.* National Bureau of Economic Research. http://www.nber.org/chapters/c2124. Accessed 25 May 2020.

Entele, B. R. (2020). Analysis of households' willingness to pay for a renewable source of electricity service connection: Evidence from a double-bounded dichotomous choice survey in rural Ethiopia. *Heliyon, 6*, 1–13.

EPA. (2011). *Assessing the Multiple Benefits of Clean Energy* (Report EPA-430-R-11-014). Environmental Protection Agency.

Eriksson, O. (2017). Nuclear power and resource efficiency—A proposal for a revised primary energy factor. *Sustainability, 9*, 1063. https://doi.org/10.3390/su9061063

Etemad, B., & Luciani, J. (1991). *World energy production 1800–1985*. DROZ.

Etzkowitz, H., & Leydesdorff, L. (1995). The triple helix – University-industry-government relations: A laboratory for knowledge based economic development. *EASST Review, 14*(1), 14–19.

European Commission. (2020). *2020 report on the State of the Energy Union pursuant to Regulation (EU) 2018/1999 on Governance of the Energy Union and Climate Action*, Brussels, 14.10.2020 COM(2020) 950 final.

Eurostat. (2021). http://ec.europa.eu/eurostat/data/database. Accessed 15 May 2021.

Eyer, J., & Corey, G. (2010). *Energy storage for the electricity grid: Benefits and market potential assessment guide*, SAND2010-0815, Albuquerque, Sandia National Laboratories.

FAO. (2001). *State of the world 2001*. Climate change. http://www.fao.org/3/y0900e/y0900e06.htm#:~:text=Carbon%20sequestration&text=Typical%20sequestration%20rates%20for%20afforestation,et%20al.%2C%201996. Assessed 05 Jan 2021.

FAO. (2012). *State of the world's forest* (Chapter 2) (p. 9). FAO.

FAO. (2016). *National statistics related to woodfuel and international recommendations*. http://www.fao.org/3/ca6504en/ca6504en.pdf. Accessed 10 Jan 2021.

FAO, UNEP. (2020). *The state of the world's forests 2020. Forests, biodiversity and people,* Rome. https://doi.org/10.4060/ca8642en. Accessed 10 July 2021.

Farhar B. C. (1999). *Willingness to pay for electricity from renewable resources: A review of utility market research.* NREL, US Department of Energy.

Fasol, K. H. (2002). A short history of hydropower control. *IEEE Control System Magazine*, 68–76.

Fattouh, B., Poudineh, R., & West, R. (2019). The rise of renewables and energy transition: What adaptation strategy exists for oil companies and oil-exporting countries? *Energy Transitions, 3*, 45–58.

Fell, M. J. (2017). Energy services: A conceptual review. *Energy Research & Social Science, 27*, 129–140.

Ferguson, J. (2021). *Europe's love affair with fossil fuels*. https://www.investigate-europe.eu/en/2020/europes-love-affair-with-fossil-fuels/. Accessed 20 Apr 2021.

Forbes. (2020). https://www.forbes.com/sites/tomzeller/2015/04/21/natural-gas-leaks-a-30-billion-opportunity-and-global-warming-menace/. Accessed 10 Mar 2020.

Fouquet, R. (2011). Divergences in long run trends in the prices of energy and energy services. *Review of Environmental Economics and Policy, 5*(2), 196–218.

Fouquet, R. (2016). Historical energy transitions: Speed, prices and system transformation. *Energy Research & Social Science, 22*, 7–12. ISSN 2214-6296.

FuelCellsWorks. (2021). Massive Green Hydrogen Production Project will soon be unveiled in Bécancourt, H2 V Energies, Québec. https://fuelcellsworks.com/news/h2-v-energies-massive-green-hydrogen-production-project-will-soon-be-unveiled-in-becancour-quebec/. Accessed 5 July 2021.

Gales, B., Kander, A., Malanima, P., & Rubio, M. (2007). North versus South: Energy transition and energy intensity in Europe over 200 years. *European Review of Economic History, 11*(2), 219–253.

Gasparatos, A., Borzoni, M., & Abramovay, R. (2012). The Brazilian bioethanol and biodiesel programs: Drivers, policies and impacts. In A. Gasparatos & P. Stromberg (Eds.), *Socio-economic and environmental impacts of biofuels*. Cambridge University Press.

Georgescu-Roegen, N. (1971). *The entropy law and the economic process.* A Harvard Paperback. https://www.scribd.com/document/267519123/The-Entropy-Law-and-the-Economic-Process-by-Nicholas-Georgescu-Roegen-pdf. Accessed 3 May 2020.

Ghosh, T. K., & Prelas, M. A. (2009). *Energy Resources and Systems, Volume 1: Fundamentals of non-renewable energy* (pp. 649–676). Springer.

Ghosh, T. K., & Prelas, M. A. (2011). *Energy resources and systems, Volume 2: Renewable energy* (Vol. II). Springer.

Gianfreda, A., Parisio, L., & Pelagatti, M. (2016). Revisiting long-run relations in power markets with high RES penetration. *Energy Policy, 94*, 432–445.

Gibney, E. (2021). Fuel for world's largest fusion reactor ITER is set for run. *Nature,* 22-2-2021. https://www.nature.com/articles/d41586-021-00408-1

Gielen, D., Boshell, F., Saygin, D., Bazilian, M. D., Wagner, N., & Gorini, R. (2019). The role of renewable energy in the global energy transformation. *Energy Strategy Reviews, 24*, 38–50.

Gillingham, K., & Stock, J. H. (2018). The cost of reducing greenhouse gas emissions. *Journal of Economic Perspectives, 32*(4), 53–72.

Goldenberg, J., & Reddy, A. K. N. (1990). Energy for the developing world. *Scientific American, 263*, 111–118.

Gordon, R. J. (1990). *The measurement of durable goods prices*. University of Chicago Press.

Griffith, S., & Calish, S. (2020). *Mobilizing for a zero carbon America: Jobs, jobs, jobs, and more jobs, rewiring America.* https://www.rewiringamerica.org/jobs-report. Accessed 28 Dec 2020.

Griliches, Z. (1996). The discovery of the residual: A historical note. *Journal of Economic Literature, XXXIV*, 1324–1330.

Grilli, G. (2018). Renewable energy and willingness to pay: Evidence from a meta-analysis. *Economics and Policy of Energy and the Environment, 19*, 253–271.

Grübler, A. (2012). Energy transitions research: Insights and cautionary tales. *Energy Policy, 50*, 8–16.

Grübler, A., & Nakicenovic, N. (1996). Decarbonizing the global energy system. *Technological Forecasting and Social Change, 53*(1), 97–110.

Grübler, A., Johansson, T. B., Mundaca, J., Nakicenovic, N., Pachauri, A., Riahi, K., Rogner, H.-H., & Strupeit, L. (2005). *Global energy assessment, energy primer.* IIASA, Laxenburg. https://iiasa.ac.at/web/home/research/Flagship-Projects/Global-Energy-Assessment/GEA_Chapter1_primer_hires.pdf. Accessed 27 May 2020.

Grübler, A., Wilson, C., & Nemet, G. (2016). Appels, oranges and consistent comparison of temporary dynamics of energy transition. *Energy Policy, 22*, 18–25.

Gupta, A. K., & Hall, C. A. S. (2011). A review of the past and current state of EROI data. *Sustainability, 3*, 1796–1809.

Guruswamy, L. (2011). Energy poverty. *Annual Review of Environment and Resources, 36*, 139–161. https://doi.org/10.1146/annurev-environ-040610-090118. Accessed 11 July 2021.

Haberl, H., Erb, K. H., Krausmann, F., Gaube, V., Bondeau, A., Plutzar, C., Gingrich, S., Lucht, W., & Fischer-Kowalski, M. (2007). Quantifying and mapping the human appropriation of net primary production in earth's terrestrial ecosystems. *PNAS, 104*(31), 12942–12947.

Haldane, J. B. S. (1923, February 4). *DAEDALUS or science and the future*. Paper read to the Heretics Society, Cambridge.

Hall, C. A. S. (2017). Will EROI be the primary determinant of our economic future? The view of the natural scientist versus the economist. *Joule, 1*(20), 635–638.

Hall, C. A. S., Lambert, J. G., & Balogh, S. B. (2014). EROI of different fuels and the implications for society. *Energy Policy, 64*, 141–152.

Harrison, K.W., Remick, R., & Martin, G. D. (2010). *Hydrogen production: Fundamentals and case study summaries*. National Renewable Energy Laboratory, NREL/CP-550-47302 88, Oak Ridge.

Hecht, J. (2010). *Short history of laser development.* SPIE. https://www.spiedigitallibrary.org/journals/Short-history-of-laser-development/volume-49/issue-09/091002/Short-history-of-laser-development/10.1117/1.3483597.full?webSyncID=e46e9e6e-c7a4-9dab-6a0c-bad059329ad8&sessionGUID=83c9d902-bc99-93ce-d268-bead49a28531&SSO=1. Accessed 11 Jan 2021.

Helpman, E. (2004). *The mystery of economic growth* (1st ed.). The Belknap press of the Harvard University Press.

Herman, R., Ardekanin, A. A., & Ausubel, J. H. (1989). Dematerialization. In J. H. Ausubel & H. E. Sladovich (Eds.), *Technology and environment* (1st ed., pp. 50–69). National Academy Press.

Hirth, L. (2013). The market value of variable renewable, the effect of solar wind power variability on their relative price. *Energy Economics, 38*, 218–236.
Horizon2020. (2021). https://ec.europa.eu/programmes/horizon2020/sites/horizon2020/files/h2020_threeyearson_a4_horizontal_2018_web.pdf. Accessed 12 Feb 2021.
Horses. (2021). There were 58 million horses and twenty times more cars by 2000s. https://www.google.com/search?q=number+of+horses+worldwide&oq=number+of+horses%2C+world&aqs=chrome.1.69i57j0l5.14907j0j8&sourceid=chrome&ie=UTF-8. Accessed 10 July 2021.
Howard, R. W., & Jacobson, M. Z. (2021). How green is blue hydrogen. *Energy Science & Engineering*. https://doi.org/10.1002/ese3.956
IEA. (2014). *Africa energy outlook*. International Energy Agency.
IEA. (2015). *Estimates for 2015 are based on database of the International Energy Agency.*
IEA. (2017). http://www.iea.org/bookshop/729-CO2_Emission_from_Fuel_Combustion. Accessed 26 July 2017.
IEA. (2019). *CO2 emissions from fuel combustion* (p. 147). International Energy Agency.
IEA. (2020). https://www.iea.org/reports/energy-access-outlook-2017. Accessed 13 Nov 2020.
IEA. (2021a). https://www.iea.org/classicstats/statisticssearch/. Accessed 10 July 2021.
IEA. (2021b). https://www.iea.org/reports/world-energy-investment-2020/key-findings. Accessed 3 Apr 2021.
IEA batteries. (2021). https://www.iea.org/data-and-statistics/charts/evolution-of-li-ion-battery-price-1995-2019. Accessed 20 May 2021.
IEA EV. (2021). https://www.iea.org/data-and-statistics/charts/evolution-of-li-ion-battery-price-1995-2019
IEA grid. (2021). https://www.iea.org/fuels-and-technologies/smart-grids. Accessed 22 May 2021.
IEA materials. (2021). *The role of critical minerals in clean energy transition.* World Energy Outlook Special Report, Paris.
IEA R&D. (2005). https://www.iea.org/subscribe-to-data-services/energy-technology-rdd. Accessed 9 July 2021.
IEA RD&D. (2021). https://www.iea.org/subscribe-to-data-services/energy-technology-rdd. Accessed 2 Mar 2021.
IEA storage. (2021). *Energy storage.* https://www.iea.org/reports/energy-storage. Accessed 20 May 2021.
IEA subsidies. (2021). https://www.iea.org/topics/energy-subsidies. Accessed 3 Apr 2021.
IEA-ETSAP, IRENA. (2013). *Thermal energy storage, technology brief.* https://www.irena.org/publications/2013/Jan/Thermal-energy-storage. Accessed 24 May 2021.
IEG. (2008). *The welfare impact of rural electrification: A reassessment of the costs and benefits.* Independent Evaluation Group, World Bank.
IHA. (2020). International Hydropower Association. *A brief history of hydropower.* https://www.hydropower.org/a-brief-history-of-hydropower. Accessed 4 Apr 2020.
Interest rate. (2020). *Interest rates from 1694 to 2013.* https://docs.google.com/spreadsheets/d/1OKo38R1blO71SGNIVuhaRZ4P2GYPpJQklmOymQiXixQ/edit?hl=en&hl=en#gid=0. Accessed 22 May 2020.
Interreg. (2021). *Renewable energy communities policy brief interreg Europe.* https://www.interregeurope.eu/fileadmin/user_upload/plp_uploads/policy_briefs/2018-08-30_Policy_brief_Renewable_Energy_Communities_PB_TO4_final.pdf. Accessed 9 July 2021.
Invest Europe. (2019). *Investing in Europe, private equity activity 2019*, Brussels.
IPCC. (2012). *Change, renewable energy sources and climate change mitigation* (Intergovernmental, Panel on Climate) (p. 19). Cambridge University Press.
IRENA. (2017). *Electricity storage and renewables: Costs and markets to 2030.* International Renewable Energy Agency.

IRENA. (2020). https://www.irena.org/publications/2020/Jul/Renewable-energy-statistics-2020. Accessed 7 Mar 2021.

IRENA. (2021a). https://www.irena.org/newsroom/pressreleases/2020/Nov/Tripling-Renewables-Investment-to-Reach-Climate-Goal. Accessed 3 Apr 2021.

IRENA. (2021b). https://www.irena.org/Statistics. Accessed 2 Apr 2021.

IRENA hydrogen. (2020). *Green hydrogen cost reduction: Scaling up electrolysers to meet the 1.5^{0}C climate goal*. International Renewable Energy Agency.

IRENA, IEA and REN21. (2018). *Renewable energy policies in a time of transition*. IRENA, OECD/IEA and REN21.

IRENA Jobs. (2020). *Renewable energy and jobs, annual review 2020*, Abu Dhabi.

IRENA secretariate. (2012). *Hydropower; renewable energy technologies*, Cost analysis series, vol 1(3/5).

Iszak, K., & Edler, J. (2011). *Trends and challenges in demand-side innovation policies*. Technopolis group, mimeo.

Jackson, T. (2011). *Prosperity without growth* (2nd ed.). Earthscan.

Jacobsson, S., & Bergkek, A. (2004). Transforming the energy sector: The evolution of technological systems in renewable energy technologies. *Industrial and Corporate Change, 13*(5), 815–849.

Jessel, S., Sawyer, S., & Hernandez, D. (2019). Energy, poverty, and health in climate change: A comprehensive review of an emerging literature. *Frontiers Public Health, 12* December 2019. https://doi.org/10.3389/fpubh.2019.00357. Accessed 11 July 2021.

JRC. (2012). *Analysis of energy saving potentials in energy generation*. Joint Research Center.

Jurich, K. (2016). *CO2 emission factors for Fossil Fuels*. Umwelt Bundesambt, report 28, Berlin.

Kaiser, B. A. (2020). *The transition from whale oil to petroleum*. University of South Denmark. https://www.sdu.dk/-/media/files/om_sdu/institutter/miljo/energycenter/brooks+kaiser.pdf. Accessed 17 Feb 2020

Kalimeris, P., Richardson, C., & Bithas, K. (2014). A meta-analysis investigation of the direction of the energy-GDP causal relationship: implications for the growth-degrowth dialogue. *Journal of Cleaner Production, 67*, 1–13.

Kamien, M. I., & Schwarz, N. L. (1982). *Market structure and innovation*. Cambridge Publishers.

Kammertlander, A., & Schultze, G. G. (2020). Are democracies cleaner? *European Journal of Political Economy, 64*, 101920.

Kamp, L. M. (2006). Danish and Dutch wind energy policy 1970–2000: Lessons for the future. *International Journal of Environment and Sustainable Development, 5*(2), 213–220.

Kaplan, S. N., Sensoy, B. A., & Strömberg, P. (2009). Should investors bet on the jockey or the horse? Evidence from the evolution of the firms from early business plan to public companies. *The Journal of Finance, LXIV*(1), 75–115.

Keen, M., Parry, I., & Davies, M. (2019). *Fiscal policies for Paris climate change strategies—From principle to practices*. International Monetary Fund.

Kim, J., Park, J., Kim, J., & Heo, E. (2013). Renewable electricity as a differentiated good? The case of the Republic of Korea. *Energy Policy, 54*, 327–334.

Kirzner, I. (1997). Entrepreneurial discovery and the competitive market process: An Austrian approach. *Journal of Economic Literature, 35*(1), 60–85.

Klein, A., Pfluger, B., Held, A., Ragwitz, M., Resch, G., & Faber, T. (2008). *Evaluation of different feed-in tariff design options – Best practice paper for the international feed-in cooperation*. Fraunhofer Institut, Institut for Systems and Innovation Research.

Kodama, F. (1995). *Emerging patterns of innovation; ; Sources of Japan's technological edge*. Harvard Business School Press.

Koornneef, J., Junginger, M., & Faaij, A. (2007). Development of fluidized bed combustion—An overview of trends, performance and cost. *Progress in Energy and Combustion Science, 33*, 19–55.

Kotler, P., & Zaltman, G. (1971). Social marketing: An approach to planned social change. *Journal of Marketing, 35*, 3–12.

Krausmann, F., Gingrich, S., Eisenmenger, N., Erb, K.-H., Haberl, H., & Fischer-Kowalski, M. (2009). Growth in global materials use, GDP and population during the 20th century. *Ecological Economics, 68*, 2696–2705.

Krepelkova, M. (2017). *Evolution of batteries: From experiments to everyday usage*. https://www.google.com/search?ei=Gbb8X9bhJ4WakwX0i6rwAg&q=history+of+batteries+pdf&oq=history+of+batteries&gs_lcp=CgZwc3ktYWIQARgBMgUIABDJAzICCAAyAggAMgIIADIGCAAQFhAeMgYIABAWEB4yBggAEBYQHjIGCAAQFhAeMgYIABAWEB4yBggAEBYQHjoECAAQRzoHCAAQyQMQQzoECAAQQzoCCC46CAgAEMkDEJECOgUIABCRAjoICC4QxwEQrwFQjAxYsixg3DloAHACeACAAUaIAdIGkgECMTSYAQCgAQGqAQdnd3Mtd2l6yAEIwAEB&sclient=psy-ab. Accessed 11 Jan 2021.

Krlev, G., Münscher, R., Mülbert, K. (2013). *Social Return on Investment (SROI): State–of-the-art and perspectives*. Centre for Social Investment, University of Heidelberg. http://www.ub.uni-heidelberg.de/archiv/18758. Accessed 31 May 2019.

Krohn, S. P.-E., & Morthorst, S. A. (2009). *The economics of wind energy*. Risoe National Laboratory.

Krozer, Y. (1987). Biomassa, product van de toekomst? In F. Hendriks (Ed.), *Shell*. SOMO/Jan van Arkel.

Krozer, Y. (2008). *Innovations and the environment*. Springer.

Krozer, Y. (2012). Renewable energy in European regions. *International Journal of Innovation and Regional Development, 4*(1).

Krozer, Y. (2015). *Theories and practices on innovations for sustainable development*. Springer.

Krozer, Y. (2017). Energy initiatives in Europe. In I. Visa & A. Duta (Eds.), *Nearly zero energy communities*. Springer.

Krozer, Y. (2019). Valorisation of energy services: Essay on the value addition due to renewable energy. *Energy, Sustainability and Society, 9*(9), 1–16.

Krozer, Y., & Nentjes, A. (2006). An essay on innovation for sustainable development. *Journal on Intergrative Environmental Science, 3*, 163–174.

Krozer, Y., & Vis, J. C. (1998). How to get LCA in the right direction? *Journal of Cleaner Production, 6*(1), 53–61.

Krozer, Y., Coenen, F., Hanganu, J., Lordkipanidze, M., & Sbarcea, M. (2020). Towards innovative governance of nature areas. *Sustainability, 122*, 624.

Krzywiec, P. (2018). The birth and development of oil and gas industry in the Northern Carpathian. In J. Graig, F. Gerali, F. MacAulay, & R. Sorkhabi (Eds.), *History of the European oil and gas industry* (Special Publication 465) (pp. 165–189). Geological Society.

Kuipers, S. K., & Nentjes, A. (1973). Pollution in a neo-classical world: The classics rehabilitated? *The Economist, 121*(1), 52–67.

Labandeira, X., Labeaga, J. M., & Lopez-Otero, X. (2015). *A meta-analysis on the price elasticity of energy demand*. University de Vigo, WP04/2015.

Lagarde, C., & Gaspar, V. (2019). *Getting real on meeting Paris climate change commitments*. IMF. https://blogs.imf.org/2019/05/03/getting-real-on-meeting-paris-climate-change-commitments/?utm_medium=email&utm_source=govdelivery. Accessed 14 Dec 2020.

Lancaster, K. J. (1966). A new approach to consumer theory. *Journal of Political Economy, 74*(2), 132–157.

Lander, D. S. (1998). *The wealth and poverty of nations*. W.W. Norton & Co.

Layton, B. E. (2008). A comparison of energy densities of prevalent energy sources in units of joules per cubic meter. *International Journal of Green Energy*, November.

Lazard. (2020). *Lazard levelized cost of energy analysis- version 13*. November 2019.

Lenzen, M. (2008). Life cycle energy and greenhouse gas emissions of nuclear energy: A review. *Energy Conversion and Management*, 49, 2178–2199.

Levin, T., & Thomas, V. M. (2016). Can developing countries leapfrog the centralized electrification paradigm? *Energy for Sustainable Development, 31*, 97–107.

Li, P., Peng, C., Wang, M., Li, W., Zhao, P., Wang, K., Yang, Y., & Zhu, Q. (2017). Quantification of the response of global terrestrial net primary production to multifactor global change, *Ecological Indicators*, 76, 245–255. https://www.sciencedirect.com/science/article/abs/pii/S1470160X17300274. Accessed 11 July 2021.

Lovering, J. R., Yip, A., & Nordhaus, T. (2016). Historical construction costs of global nuclear powers reactors. *Energy Policy, 91*, 371–382.

Ma, C., Rogers, A. A., Kragt, M. E., Zhang, F., Polyakov, M., Chalak, M., Pandit, R., & Tapsuwan, S. (2015). Consumers' willingness to pay for renewable energy: A meta-regression analysis. *Resource and Energy Economics, 42*, 93–109.

MacKay, D. J. C. (2009). *Sustainable energy without hot air*. UIT.

Maddison, A. (2007). Measuring and interpreting world economic performance 1500–2001. *Review of Income and Wealth, 51*(1), 1–35.

Maddison project database, version. (2018). Jutta, B., Inklaar, R., de Jong, H., & Luiten van Zanden J. (2018). *Rebasing 'Maddison': new income comparisons and the shape of long-run economic development*. Maddison Project Working Paper, nr. 10. Available for download at www.ggdc.net/maddison. Accessed 21 Mar 2020.

Malamina, P. (2016). Energy consumption in England and Italy, 1560–1913.Two pathways toward energy transition. *Economic History Review, 69*(1), 78–103.

Malanima, P. (2014). Energy in history. In N. Agnoletti & S. Neri Serneri (Eds.), *The basic environmental history*. Springer.

Malthus, T.R. ((1798) 1826). *An essay on the principle of population* (Vol. 1, 1826, 6th ed.) https://oll.libertyfund.org/titles/malthus-an-essay-on-the-principle-of-population-vol-1-1826-6th-ed. Accessed 2 May 2020.

Markandya, A., & Wilkinson, P. (2007). Electricity generation and health. *Lancet.com*, Vol. 370 September 15, 2007.

Masdar. (2021). https://grist.org/climate-energy/the-worlds-first-zero-carbon-city-is-a-big-failure/. Accessed 11 July 2021.

Mazzucato, M. (2018). Mission-oriented innovation policies: Challenges and opportunities. *Industrial and Corporate Change, 27*(5), 803–815.

Mazzucato, M., & Semieniuk, G. (2017). Public financing of innovation: New questions. *Oxford Review of Economic Policy, 33*(1), 24–48.

McDonald, A., & Schrattenholzer, L. (2003). *Learning curves and technology assessment*. International Institute for Applied Systems Analysis.

Meadows, D. H., Meadows, D. L., Randers, J., & Behrens III, W. W. (1972). *Limits to growth*. A Potomac Associates Book. http://donellameadows.org/the-limits-to-growth-now-available-to-read-online/. Accessed 2 May 2020.

Mehmeti, A., Angelis-Dimakis, A., Arampatzis, G., McPhail, S. J., & Ulgiati, S. (2018). Life cycle assessment and water footprint of hydrogen production methods: From conventional to emerging technologies. *Environments, 5*, 1–19.

Melsted, O., & Palua, I. (2018). The historical transition from coal to hydrocarbons; previous explanations and the need for an integrative perspective. *Canadian Journal of History, 53*(3), 395–423.

Mercure, J.-F., Pollitt, H., Chewpreecha, U., Salas, P., Foley, A. M., Holden, P. B., & Edwards, N. R. (2014). The dynamics of technology diffusion and the impacts of climate policy instruments in the decarbonisation of the global electricity sector. *Energy Policy, 73*, 686–700.

Metz, B., Davidson, O., de Coninck, H., Loos, M., & Meyer, L. (2005). *Carbon dioxide capture and storage* (Intergovernmental Panel on Climate Change) (pp. 43–44). Cambridge University Press.

Milanovic, B. (2012). *Global income inequality by the numbers: In history and now*. The World Bank, report nr. 6259.

Miranowski, J. (2005). Energy consumption in US agriculture. In J. Outlaw, K. Collins, & J. Duffield (Eds.), *Agriculture as a producer and consumer of energy* (pp. 68–111). CAB.

Mitrova, T. (2019). *Global and Russian energy outlook*. Skolkovo Energy Centre, Moscow School of Management.

Mokyr, J. (2005). *The gift of Athene* (Chapter 2). Princeton University Press.

Murkin, C., & Brightling, J. (2016). Eighty years of steam reforming. *The Johnson Matthey Technology Review, 60*(4), 263–269.

Nakicenovic, N. (2012). *Global energy assessment – Toward a sustainable future.* Cambridge University Press/The International Institute for Applied Systems Analysis.
Narodoslawski, M. (2019). Bioresources and technologies. In Y. Krozer & M. Narodoslawski (Eds.), *Economics of bioresources.* Springer.
NaturalGas.org. (2020). http://naturalgas.org/overview/history/. Accessed 4 Apr 2020.
Negro, S. O., Alkemade, F., & Hekkert, M. P. (2012). Why does renewable energy diffuse so slowly, a review of innovation system problems. *Renewable and Sustainable Energy Reviews, 16*, 3836–3846.
Nelson, R. (1995). Recent evolutionary theorizing about economic change. *Journal of Economic Literature, XXXIII*, 48–90.
Neubauer J., & Simpson, M. (2015). *Deployment of behind-the- meter energy storage for demand charge reduction.* NREL/TP-5400-63162, Golden.
Nordhaus, W. (1973). The allocation of energy resources. *Brookings Papers on Economic Activity, 3.* http://www.brookings.edu/~/media/projects/bpea/1973%203/1973c_bpea_nordhaus_houthakker_solow.pdf. Visited, 20 Nov 2014.
Nordhaus, W. D. (1994). *Do real output and real wages measures capture realities? The history of lighting suggests not.* Cowles Foundation.
Nordhaus, W. D. (2010). *The perils of learning model for modelling endogenous technical change.* https://www.nap.edu/read/13023/chapter/15. Accessed 25 May 2020.
NREL. (1997). *Dollars from sense, the economic benefits of renewable energy.* National Renewable Energy Laboratory.
NREL. (2020). *Cell-efficiency.* https://www.nrel.gov/pv/cell-efficiency.html. Accessed 20 Apr 2020.
NREL. (2021a). *Levelized cost of energy calculator.* https://www.nrel.gov/analysis/tech-lcoe.html. Accessed 17 Apr 2021.
NREL. (2021b). https://www.nrel.gov/analysis/tech-lcoe-re-cost-est.html. Accessed 17 Apr 2021.
Ntanos, S., Kyriakopoulos, G., Chalikias, M., Arabatzis, G., & Skordoulis, M. (2018). Public perceptions and willingness to pay for renewable energy: A case study from Greece. *Sustainability, 10*(687), 1–16.
Nuvolari, A. (2010). The theory and practice of steam engineering in Britain and France, 1800–1850, *Documents pour l'histoire des techniques*, 19, p. 189–197. https://journals.openedition.org/dht/1439. Accessed 10 July 2021.
O'Connor, P. A., & Cleveland, C. J. (2014). U.S. energy transition 1780–2010. *Energies, 7*, 7955–7993.
O'Mahony, T. (2020). State of the art in carbon taxes: A review of the global conclusions. *Green Finance, 2*(4), 409–423.
OECD. (2021). https://stats.oecd.org/OECDStat_Metadata/ShowMetadata.ashx?Dataset=RE_FIT&ShowOnWeb=true&Lang=en. Accessed 3 Apr 2021.
OECD RD&D. (2021). *Total RD&D, Group 1. Energy efficiency.* https://stats.oecd.org/. Accessed 17 May 2021.
Ornetzeder, M., & Rohracher, H. (2006). User-led innovations and participation processes: Lessons from sustainable energy technologies. *Energy Policy, 34*, 138–150.
Ortiz-Rivera, E. I., Reyes-Hernandez, A. L., & Febo, R. A. (2007). *Understanding the history of fuel cells.* IEEE. https://ieeexplore.ieee.org/document/4510259. Accessed 9 July 2021.
Pandyaswargo, A. H., Jagath, P., Gamaralalage, D., Chen, L., Knaus, M., Onoda, H., Mahichi, F., & Yanghui, G. (2019). Challenges and an implementation framework for sustainable municipal organic waste management using biogas technology in emerging asian countries. *Sustainability, 11*(6631), 1–27.
Papaioannou, K. J., & Luiten van Zanten, J. (2015). The dictator effect: How long years in office affect economic development. *Journal of Institutional Economics, 11*(1), 111–139.
Pepermans, G., Driesen, J., Haeseldonckx, D., D'haeseleer, W., & Belmans, R. (2003). *Distributed generation, definition, benefits and issues.* University of Leuven.
Pernick, R., & Wilder, C. (2008). *The cleantech revolution.* Harper.

Perrin, B., (2021), How to – and how not to – evaluate innovation. http://www.mande.co.uk/docs/perrin.htm. Accessed 9 July 2021.
Petrovsky, H. (1994). *The evolution of useful things* (1st ed.). Vintage Books.
Petrovsky, H. (1996). *Invention by design*. Harvard University Press.
Pieprzyk, B., & Hilje, P. R. (2015). *Flaring and venting of associated gas, current development and impacts of marginal oils*. Energy Research Architecture and https://blogs.worldbank.org/opendata/global-gas-flaring-inches-higher-first-time-five-years. Accessed 10 Mar 2020.
Piketty, T. (2014). *Capital in the twenty-first century* (1st ed.). The Belknap Press the Harvard University Press.
Pinker, S. (2018). *Enlightment now, the case for reason. Science, humanism and progress*. Penguin Books, Random House.
Plumer, B. (2016). *Solar power is contagious. This map shows how it spreads*. https://www.vox.com/2016/5/4/11590396/solar-power-contagious-maps. Accessed 18 Nov 2020.
Podobnik, B. (2006). *Global energy shifts*. TERI.
Pommeranz, K. (2000). *The great divergence*. Princetown University Press.
Popkin, G. (2019). How much can forests fight climate change? *Nature, 565*, 280–282. https://doi.org/10.1038/d41586-019-00122-z
PWC. (2017). *Effectiveness of tax incentives for venture capital and business angels to foster the investment of SMEs and start-ups* (European Commission, Working Paper 68-2017).
Pyper, J. (2017). *US advanced energy revenue grew by just 1% in 2016*, GTM, 7-3-2017. https://www.greentechmedia.com/articles/read/u.s.-advanced-energy-revenue-grew-by-just-1-in-2016. Accessed 17 May 2021.
Quaschning, V. (2008). *Renewable energy and climate change*. Carl Hanser Verlag.
Ramirez, C. A., & Worrell, E. (2006). Feeding fossil fuels to the soil. An analysis of energy embedded and technological learning in the fertilizer industry. *Resources, Conservation and Recycling, 46*, 75–93.
Rapier, R. (2020). Life cycle emission of hydrogen. *Climate* 5/6/2020. https://4thgeneration.energy/life-cycles-emissions-of-hydrogen/. Accessed 14 May 2021.
Raugeri, M., & Leccisi, E. (2016). A comprehensive assessment of the energy performance of the full range of electricity generation technologies deployed in the United Kingdom. *Energy Policy, 90*, 46–59.
Raworth, K. (2017). Donut Economie, *Nieuw Amsterdam*, Amsterdam (Dutch translation).
REFINITIV. (2021). *Carbon market year in review*, Oslo, Norway.
Reichmuth, D., & Goldman J. (2017). *Accelerating US leadership in electric vehicles union of concerned scientists*. Policy Brief, August, 2017. https://www.ucsusa.org/clean-vehicles/electric-vehicles/accelerating-us-leadership-electric-vehicles-2017. Accessed 11 Feb 2019.
Reij, C., & Winterbottom, R. (2015). *Scaling up regreening: Six steps to success*. World Resource Institute. https://www.wri.org/insights/can-we-restore-350-million-hectares-2030#:~:text=Many%20national%20tree%2Dplanting%20campaigns,require%20some%20%2434%20billion%20annually. Accessed 23 May 2021.
Resch, G., Held, A., Faber, T., Panzer, C., Toro, F., & Haas, R. (2008). Potentials and prospects for renewable energies at global scale. *Energy Policy, 36*, 4048–4056.
Richard, K. R., & Stokes, C. (2004). A review of forest carbon sequestration cost studies: A dozen years of research. *Climatic Change, 63*, 1–48.
Richter, M. (2012). Utilities' business models for renewable energy: A review. *Renewable and Sustainable Energy Reviews, 16*, 2483–2493.
Rismanchi, B. (2017). District energy networks (DEN), current global status and future development. *Renewable and Sustainable Energy Reviews, 75*, 571–579.
Ritchie, H., & Roser, M. (2020). *Energy, Published online at OurWorldInData.org*. Retrieved from: https://ourworldindata.org/energy [Online Resource]. Accessed 21 Mar 2020.
Roberts, D. (2019). Getting to 100% renewables requires cheap energy storage. But how cheap? *VOX*, 20-9-2019. https://www.vox.com/energy-and-environment/2019/8/9/20767886/

renewable-energy-storage-cost-electricity#:~:text=To%20provide%20baseload%2C%20intermediate%2C%20bipeaker,%2410%E2%80%9330%2FkWh%20. Accessed 11 July 2021.
Rockström, J., Steffen, W., Noone, K., Persson, Å., Chapin, F. S., Lambin, E., Lenton, R. M., Scheffer, M., Folke, C., Schellnhuber, H., Nykvist, B. C. A., de Wit, A., Hughes, T., van Leeuw, S., Rodhe, H., Sörlin, S., Snyder, P. K., Costanza, R., Svedin, U., . . . Foley, J. (2009). Planetary boundaries: Exploring the safe operating space for humanity. *Ecology and Society, 14*(2), 32. [online] URL: http://www.ecologyandsociety.org/vol14/iss2/art32/. Accessed 11 July 2021
Rogers, E. M. (1983). *Diffusion of innovations* (3rd ed.). The Free Press.
Rogner, H.-H. (2000). *Energy resources in UNDP* (p. 167). World Energy Assessment.
Rojas, J., Duprat, C., Engrand, E., Dartois, I., Delauche, M., Godard, M., Gounelle, J. D., Carrillo-Sánchez, P., & Pokorný, J. M. C. (2021). The micrometeorite flux at Dome C (Antarctica), monitoring the accretion of extraterrestrial dust on Earth. *Earth and Planetary Letters, 560*(116794).
Rosenberg, N. (1982). Learning by using. In N. Rosenberg (Ed.), *Inside the black box* (pp. 104–119). Cambridge University Press.
Rosenberg, N. (1994). Energy efficient technologies. In A. Q. Curzio, M. Fortis, & R. Zoboli (Eds.), *Innovation, resources and economic growth* (pp. 63–82). Springer.
Rosenberg, N., & Birdzell, L. E. (1986). *How the west grew rich*. I.B. Tauris & Co. Ltd, Publishers.
Rosenberg, D., & Tarasenko, G. (2020). Innovation for despots? How dictators and democratic leaders differ in stifling innovation and misusing natural resources across 114 countries. *Energy Research and Social Science, 68*, 101543.
Rubin, E. S., Azevedo, I. M. L., Jaramillo, P., & Yeh, S. (2015). A review of learning rates for electricity supply technologies. *Energy Policy, 86*, 198–218.
Ruether, J. A., Ramezan, M., & Balash, P. C. (2004). Greenhouse gas emission from coal gasification power generation system. *Journal of Infrastructure System, 10*(3), 111–119.
Sachs, J. D., & Warner, A. M. (2001). Natural resources and economic development, the curse of natural resources. *European Economic Review, 45*, 827–838.
Sarasvathy, S. (2001). Causation and effectuation: Toward a theoretical shift from economic inevitability to entrepreneurial contingency. *Academy of Management Review, 26*(2), 243–263.
Saygin, D., Worrel, E., Patel, M. K., & Gielen, D. J. (2011). Potential of best practice technologies to improve energy efficiency in the global chemical and petrochemical sector. *Energy, 36*, 5579–5790.
Scarpa, R., & Willis, K. (2010). Willingness-to-pay for renewable energy: Primary and discretionary choice of British households' for micro-generation technologies. *Energy Economics, 32*(1), 129–136.
Schiermeier, Q., Tollefson, J., Scully, T., Witze, A., & Morton, O. (2008). Energy alternatives: Electricity without carbon. *Nature, 454*, 816–823.
Schleicher-Tappeser, R. (2012). How renewables will change electricity markets in the next five years. *Energy Policy, 48*, 64–75.
Schumacher, E. F. (1973). *Small is beautiful*. Blond & Briggs.
Schumpeter, J. A. ((1939) 1989). *Business cycles* (4th ed.). Porcupine Press.
Sciencedirect. (2021). Table 3.8. https://www.sciencedirect.com/topics/engineering/hydrogen-production-cost#:~:text=Hydrogen%20Production%20and%20Current%20Technologies&text=For%20instance%2C%20the%20hydrogen%20production,of%200.3%20US%24%2Fkg. Accessed 11 July 2021.
Scientific American. (2020). https://www.scientificamerican.com/article/earthtalks-volcanoes-or-humans/ published 11-2-2009. Accessed 31 July 2020.
Seijlhouwer M. (2021). Dit zijn de grootste waterstofprojecten ter wereld. *Change Inc*. https://www.change.inc/energie/grootste-waterstof-fabriek-project-ter-wereld-36699. Accessed 5 July 2021.
Sen, A. (2002). *Rationality and freedom*. The Belknap Press of Harvard University.
Sen, A. (2009). *The idea of justice* (1st ed.). Penguin Books.

Seyfang, G., & Haxeltine, A. (2012). Growing grassroots innovations: Exploring the role of community-based initiatives in governing sustainable energy transition. *Environment and Planning C: Government and Policy, 30*, 381–400.
Shafiee, S., & Topal, E. (2009). When will fossil fuel reserves be diminished. *Energy Policy, 37*, 181–189.
Shafiee, S., & Topal, E. (2010). A long term view on fossil fuel prices. *Applied Energy, 87*, 988–1000.
Shahan, Z. (2020). *History of wind turbines, renewable energy world.* https://www.renewableenergyworld.com/2014/11/21/history-of-wind-turbines/#gref. Accessed 4 Apr 2020.
Shane, S. (2000). Prior knowledge and discovery of entrepreneurial opportunities. *Organization and Science, 11*(4), 448–469.
Siao, P. (2014). Demand response and smart grid – A survey. *Renewable and Sustainable Energy Reviews, 30*, 461–478.
Siegel, J. (2008). *Investing in renewable energy*. Wiley.
Smil, V. (2000). Energy in the twentieth century: Resources, conversions, costs, uses, and consequences. *Annual Review of Energy and the Environment, 25*, 21–51.
Smil, V. (2004). *World history and energy* (Encyclopedia of Energy, Vol. 6. r 2004). Elsevier.
Smil, V. (2016). Examining energy transitions: A dozen insights based on performance. *Energy Research & Social Science, 22*, 194–197.
Smil, V. (2017). *Energy transition*. Praeger.
Smil, V. (2019). *Energy and civilization, a history*. The MIT Press.
Solana, J. (2020). Climate change litigation as financial risk. *Green Finance, 2*(4), 344–372.
Solorio, T., & Bocquillon, P. (2017). EU renewable energy policy: A brief overview of its history and evolution. In I. Solorio & H. Jorgens (Eds.), *A guide to EU renewable energy policy* (Monograph Book) Elgar Online, https://www.elgaronline.com/view/edcoll/9781783471553/9781783471553.00011.xml. Accessed 31 Mar 2021
Solow, R. M. ((1973) 1977). The economics of resources or the resources of economics, reprint. In R. Dofman, & N. S. Dorfman (Eds.), *Economics of the environment* (1st ed., pp. 354–370). W. W. Norton & Company Inc.
Sovacool, B. K. (2012). The political economy of energy poverty: A review of key challenges. *Energy for Sustainable Development, 16*(2012), 272–282.
Sovacool, B. K. (2016). How long will it take, conceptualizing the temporary dynamics of energy transition. *Energy Research & Social Science, 13*, 202–215.
Spath, P. L., & Mann, M. K. (2001). *Life cycle assessment of hydrogen production via natural gas steam reforming*. National Renewable Energy Laboratory.
Spiesberger, M., & Schönbeck, J. (2019). Innovation vouchers for the transition of energy and innovation systems. *Foresight and STI Governance, 13*(1), 70–76.
Staffel, I., Scamman, D., Velazquez Abad, A., Balcombe, P., Dodds, P. E., Ekins, P., Shah, N., & Ward, K. R. (2018). The role of hydrogen and fuel cells in the global energy system. *Energy and Environmental Science*. https://doi.org/10.1039/c8ee01157e
Stahel, W. R. (2010). *The performance economy*. Palgrave Macmillan.
Statista. (2021). https://www.statista.com/statistics/878705/investments-by-venture-capital-in-europe/. Accessed 27 Feb 2021.
Statista electrolysers. (2021). https://www.statista.com/statistics/1179559/hydrogen-electrolyzers-by-capacity-worldwide-location/. Accessed 3 July 2021.
Statista smart grid. (2021). https://www.statista.com/statistics/222082/projected-us-smart-grid-market-size. Accessed 17 May 2021.
Statista Solar Power. (2021). https://www.statista.com/statistics/289149/revenue-solar-power-industry-united-states. Accessed 16 May 2021. Sales prices are calculated as sales divided by deliveries; the sales are assumed equivalent of revenues.
Stobaugh, R. (1988). *Innovation and competition* (1st ed.). Harvard Business School Press.
Stoneman, P. (1983). *The economic analysis of technological change*. Oxford University Press.

Streimikiene, D., Balezentis, T., Alisauskaite-Seskiene, I., Stankuniene, G., & Simanaviciene, Z. (2019). A review of willingness to pay studies for climate change mitigation in the energy sector. *Energies, 12*(1481), 1–33.

Sullivan, W. G., Wicks, E. M., & Koelling, P. (2009). *Engineering economy*. Pearson.

Suneeta, D., Fernandes, N., Trautmann, M., Streets, D. G., Roden, C. A., & Bond, T. C. (2007). Global biofuel use, 1850–2000. *Global Biogeochemical Cycles, 21*, 1–15.

Symons, J., & Karlsson, R. (2015). Green political theory in a climate changed world: between innovation and restraint. *Environmental Politics, 24*(2), 173–192.

Taleb, N. N. (2012). *Antifragile* (1st ed.). Random House.

Taylor, B. (2017). *The future of energy prices: Lessons from 750 years of history*, January, 5, 2017, blog Global Financial Data. http://www.globalfinancialdata.com/the-future-of-energy-prices-lessons-from-750-years-of-history/. Accessed 14 Dec 2020.

Taylor, M. (2020). *Energy subsidies: Evolution in the global energy transformation to 2050*. International Renewable Energy Agency.

Teske, S., Sawyer, S., & Schäfer, O. (2015). *Energy [R]evolution*. Greenpeace.

The Economist. (2015, August 8). https://www.economist.com/leaders/2015/08/08/time-to-fix-patents. Accessed 11 Feb 2021.

Thurow, L. (1980). *Zero Sum Society* (p. 197). Basic Books Inc.

Timilsina, G. R. (2020). *Demystifying the costs of electricity generation technologies* (World Bank, Policy paper 9303). Washington, DC.

Tirole, J. (2017). *Economics for the common good*. Princeton University Press.

Toffler, A. (1980). *The third wave* (1st ed.). Bantam Books.

Tol, R. S. J. (2005). Marginal damage costs of carbon dioxide emissions. *Energy Policy, 33*, 2064–2074.

Trancik, J. E. (2006). Scale and innovation in energy sector: A focus on photovoltaics and nuclear fission. *Environmental Research Letters, 1*, 1–7.

Tuballa, M. L., & Abundo, M. (2016). A review of the development of smart grids. *Renewable and Sustainable Energy Reviews, 59*, 710–725.

Tyl, B., & Lizarralde, I. (2017). The citizen funding: An alternative to finance renewable energy projects. *Procedia CIRP, 64*, 199–204.

Ulrich, O. (1979). *Weltniveau*. Ritbuch Verlag.

UN. (2020). https://www.un.org/sustainabledevelopment/. Accessed 14 Nov 2020.

UNEP/BNEF, Frankfurt School-UNEP Centre/BNEF. (2018). *Global trends in renewable energy investment 2018*. Frankfurt am Main. http://www.fs-unep-centre.org. Accessed 7 Mar 2021.

US-DOE. (2020). *Energy storage grand challenge energy storage market, technical report*, NREL/TP-5400-78461 DOE/GO-102020-5497.

Usher, B. (2019). *Renewable energy*. Columbia University Press.

van Ermen, R. (2007). *Comparison and assessment of funding schemes for development of new activities and investments in environmental technologies*. Fundetec, rapport number 044370, Brussels.

van Leeuwen, C. (1989). De organisatie van milieu en veiligheid in een grote onderneming. In H. Vollebergh (Ed.), *Milieu en Innovatie* (pp. 155–176). Wolters Noordhoff.

Velthuijsen, J.-W. (2018). *Vergelijking Elektriciteitsprijzen 2017*. PWC.

Vohra, K., Vodonos, A., Schwartz, J., Marais, E. A., Sulprizio, M. P., & Mickley, L. J. (2021). Global mortality from outdoor fine particle pollution generated by 1 fossil fuel combustion: Results from GEOS-Chem. *Environmental Research, 195*, 110754.

Voltage. (2021). https://learn.sparkfun.com/tutorials/voltage-current-resistance-and-ohms-law/all. Accessed 10 July 2021. Resistance (Ω) and voltage (V) is reciprocal to current (A) given that power in Watt (W) is current in Ampere (A) times voltage in Volt (V), its W =A x V. Ratio current to voltage depends on resistance in Ohm (Ω): $A = V/\Omega$, or $V = (W \times \Omega)^{0.5}$. It implies that high voltage is needed to overcome resistance, usually 48400 V (48.4 kV) is distributed for 220 V consumption.

von Hippel, E. (2005). *Democratisation innovations* (1st ed.). MIT Press.

von Hippel, E., de Jong, J. P. J., & Flowers, S. (2012). Comparing business and household sector innovation in consumer products: Findings from a representative study in the United Kingdom. *Management Science, 58*(9), 1669–1681.

Walker, G., & Devine-Wright, P. (2008). Community renewable energy: What should it mean? *Energy Policy, 36*, 497–500.

Wallerstein, I. (1974). *The modern world system I: Capitalist agriculture and the origins of the European world-economy in the sixteenth century*. Academic.

Warde, P. (2007). *Energy consumption in England & wales, 1560–2000*. Consiglio Nazionale delle Ricerche, Istituto di Studi sulle Società del Mediterraneo.

Watanabe, C., Griffy-Brown, C., Zhu, B., & Nagamatsu, A. (2002). Inter–firm Technology Spillover and the "Virtous Cycle" of Photovoltaic development in Japan. In A. Grübler, N. Nakicenovic, & W. D. Nordhaus (Eds.), *Technological change and the environment* (Resources for the Future) (pp. 127–159).

WEC. (2010). *Energy efficiency: A recipe for success*. World Energy Council.

WEC. (2020). *Five steps to energy storage*. World Business Council.

Wiersma, D. (1989). *De Efficiëntie van een Marktconform Milieubeleid, een uitwerking van de* SO_2 *emissiebestrijding van de Nederlandse Electriciteitssector*, Rijksuniversiteit Groningen, dissertation.

Wikipedia. (2021). *Hydrogen economy*. https://en.wikipedia.org/wiki/Hydrogen_economy. Accessed 12 May 2021.

Wilson M., Jones, J. B., & Audinet, P. (2010). *A new slant on slopes, measuring the benefits of increased electricity access in developing countries* (World Bank, Washington, Report No. 53963-GLB).

World Bank. (2016). *State and trends of carbon pricing 2016*. World Bank.

Yeh, S., & Rubin E.S. (2012). A review of uncertainties in technology experience curves. *Energy Economics, 34*, 762–771.

Ye, S., & Rubin, E. R. (2020). *Uncertainties in technology experience curves for energy – Economic models*. https://www.nap.edu/read/13023/chapter/16#88. Accessed 25 May 2020.

Yildiz, Ö. (2014). Financing renewable energy infrastructures via financial citizen participation – The case of Germany. *Renewable Energy, 68*, 677–685.

Ziegler, M. S., Mueller, J. M., Pereira, G. D., Song, J., Ferrara, M., Chiang, Y.-M., & Trancik, J. E. (2019). Storage requirements and costs of shaping renewable energy toward grid decarbonization. *Joule, 3*, 2134–2153.

Ziegler, T., Shneor, R., & Zhang, B. Z. (2020). The global status of the crowdfunding industry. In R. Shneor, L. Zhao, & B.-T. Flåten (Eds.), *Advance in crowdfunding*. Springer.

Zito, R. (2010). *Energy storage: A new approach, appendix: A history of batteries*. https://onlinelibrary.wiley.com/doi/pdf/10.1002/9780470922330.app1. Accessed 1 Jan 2021.

Zorić, J., & Hrovatin, N. (2012). Household willingness to pay for green electricity in Slovenia. *Energy Policy, 47*(2012), 180–187.